I0038461

Foundations of Microbiology

Kruger Brentt
Publishers

Foundations of Psychology

Foundations of
MICROBIOLOGY

Frances Abe
Ryan Gonçalves

Kruger Brentt
Publishers
2024

Kruger Brentt Publishers UK. LTD.
Company Number 9728962

Regd. Office: 68 St Margarets Road, Edgware, Middlesex HA8 9UU

© 2024 AUTHORS

ISBN: 978-1-78715-134-5

Disclaimer:

Every possible effort has been made to ensure that the information contained in this book is accurate at the time of going to press, and the publisher and author cannot accept responsibility for any errors or omissions, however caused. No responsibility for loss or damage occasioned to any person acting, or refraining from action, as a result of the material in this publication can be accepted by the editor, the publisher or the author. The Publisher is not associated with any product or vendor mentioned in the book. The contents of this work are intended to further general scientific research, understanding and discussion only. Readers should consult with a specialist where appropriate.

Every effort has been made to trace the owners of copyright material used in this book, if any. The author and the publisher will be grateful for any omission brought to their notice for acknowledgement in the future editions of the book.

All Rights reserved under International Copyright Conventions. No part of this publication may be reproduced, stored in a retrieval system, or transmitted in any form or by any means, electronic, mechanical, photocopying, recording or otherwise without the prior written consent of the publisher and the copyright owner.

For information on all our publications visit our website at http://krugerbrentt.com/

Preface

Microbiology is a branch of biology that deals with the study of microorganisms, including bacteria, viruses, fungi, and protozoa. Microorganisms are essential to many aspects of life, including food production, medicine, and biotechnology. Understanding the foundations of microbiology is therefore essential for developing solutions to various challenges facing society. One of the key principles of microbiology is that microorganisms are ubiquitous and can be found in almost every environment on Earth. This includes extreme environments such as hot springs and polar ice caps, as well as inside the human body. In fact, the human body contains more microbial cells than human cells, highlighting the importance of microorganisms to human health. Another important principle of microbiology is that microorganisms play important roles in many biological processes, including nutrient cycling, decomposition, and fermentation. Microorganisms are also used in the production of many foods, such as cheese, bread, and yogurt, as well as in the production of various pharmaceuticals and other biotechnological products.

Microorganisms can also cause diseases in humans, animals, and plants. Understanding the mechanisms by which microorganisms cause disease is essential for developing treatments and vaccines to prevent or treat these diseases. Microbiology also plays a critical role in the field of environmental science, where it is used to monitor and control the spread of pathogens and other contaminants in the environment. The study of microbiology involves a range of techniques and methods, including microscopy, culturing, and molecular biology techniques. Microscopy is used to observe and study the morphology and behavior of microorganisms, while culturing is used to isolate and grow microorganisms for further study. Molecular biology techniques, such as polymerase chain reaction (PCR) and gene sequencing, are used to study the genetic makeup of microorganisms and their interactions with their environment. The foundations of microbiology are essential for understanding the role of microorganisms in various aspects of life, including food production, medicine, and biotechnology. The study of microbiology is also critical for understanding the mechanisms by which microorganisms cause disease and for developing treatments and vaccines to prevent or treat these diseases. Advances in microbiology have the potential to impact many areas of society, making it an important field of study for the future.

The present book contains 23 chapters namely microbiology: an introduction; introduction to microscopes; structure of cells; cell walls of bacteria; internal components of bacteria; surface structures of bacteria; introduction to archaea; viruses: an introduction; microbial development; environmental factors; nutrition for microbes; energetics & redox reactions; introduction to chemoorganotrophy; chemolithotrophy and nitrogen metabolism; introduction to phototrophy; taxonomy and evolution; introduction to microbial genetics; introduction to genetic engineering; introduction to genomics; the symbioses of microbes; introduction to bacterial pathogenicity & introduction to viruses. The text concentrates on the essential aspects of microbiology and thus provides the student with a broadly based overview of the subject. The diagrams are neatly drawn and suitably labelled for self-understanding and the contents are dealt with in a lucid style. The content is carefully prepared incorporating all the relevant points and the subject content is so simple that the student can easily learn and remember. This book caters the needs of UG students of microbiology, it is also suitable for life-science, biochemistry, botany, zoology, medicine, pharmacy and agriculture, as well as food science, biotechnology, ecology and environmental science. We are grateful to all those persons as well as various books, manuals, periodicals, magazines, journals etc. that helped in the preparation of this book. In spite of the best efforts, it is possible that some errors may have occurred into the compilation and editing of the book. Further queries, constructive suggestions and criticisms for the improvement of the book are always welcomed and shall be thankfully acknowledged.

Frances Abe

Ryan Gonçalves

Contents

Contents

01 | Microbiology: An Introduction

INTRODUCTION: AN OVERVIEW

Welcome to the wonderful world of microbiology! Yay! So. What is microbiology? If we break the word down it translates to "the study of small life," where the small life refers to microorganisms or microbes. But who are the microbes? And how small are they?

Generally microbes can be divided into two categories: the cellular microbes (or organisms) and the acellular microbes (or agents). In the cellular camp we have the bacteria, the archaea, the fungi, and the protists (a bit of a grab bag composed of algae, protozoa, slime molds, and water molds). Cellular microbes can be either unicellular, where one cell is the entire organism, or multicellular, where hundreds, thousands or even billions of cells can make up the entire organism. In the acellular camp we have the viruses and other infectious agents, such as prions and viroids.

In this textbook the focus will be on the bacteria and archaea (traditionally known as the "prokaryotes,") and the viruses and other acellular agents.

CHARACTERISTICS OF MICROBES

Obviously microbes are small. The traditional definition describes microbes as organisms or agents that are invisible to the naked eye, indicating that one needs assistance in order to see them. That assistance is typically in the form of a microscope of some type. The only problem with that definition is that there are microbes that you can see without a microscope. Not well, but you can see them. It would be easy to dismiss these organisms as non-microbes, but in all other respects they look/act/perform like other well-studied microbes (who follow the size restriction).

So, the traditional definition is modified to describe microbes as fairly simple agents/ organisms that are not highly differentiated, meaning even the multicellular microbes are composed of cells that can act independently– there is no set division of labor. If you take a giant fungus and chop half the cells off, the remaining cells will continue to function unimpeded. Versus if you chopped half my cells off, well, that would be a problem. Multicellular microbes, even if composed of billions of cells, are relatively simple in design, usually composed of branching filaments.

It is also acknowledged that research in the field of microbiology will require certain common techniques, largely related to the size of the quarry. Because microbes are so small and there are so many around, it is important to be able to isolate the one type that you are interested in. This involves methods of sterilization, to prevent unwanted contamination, and observation, to confirm that you have fully isolated the microbe that you want to study.

MICROBE SIZE

Since size is a bit of theme in microbiology, let us talk about actual measurements. How small is small? The cellular microbes are typically measured in micrometers (μm). A typical bacterial cell (let us say E. coli) is about 1 μm wide by 4 μm long. A typical protozoal cell (let us say Paramecium) is about 25 μm wide by 100 μm long. There are 1000 μm in every millimeter, so that shows why it is difficult to see most microbes without assistance. (An exception would be a multicellular microbe, such as a fungus. If you get enough cells together in one place, you can definitely see them without a microscope!)

When we talk about the acellular microbes we have to use an entirely different scale. A typical virus (let us say influenza virus) has a diameter of about 100 nanometers (nm). There are 1000 nanometers in every micrometer, so that shows why you need a more powerful microscope to see a virus. If a typical bacterium (let us pick on E.coli again) were inflated to be the size of the Statue of Liberty, a typical virus (again, influenza virus works) would be the size of an adult human, if we keep the correct proportions.

THE DISCOVERY OF MICROBES

The small size of microbes definitely hindered their discovery. It is hard to get people to believe that their skin is covered with billions of small creatures, if you cannot show it to them. "Seeing is believing," that is what I always say. Or someone says that.

In microbiology, there are two people that are given the credit for the discovery of microbes. Or at least providing the proof of their discovery, both around the same time period:

Robert Hooke (1635-1703)

Robert Hooke was a scientist who used a compound microscope, or microscope with two lenses in tandem, to observe many different objects. He made detailed drawings of his observations, publishing them in the scientific literature of the day, and is credited with publishing the first drawings of microorganisms. In 1665 he published a book by the name of Micrographia, with drawing of microbes such as fungi, as well as other organisms and cell structures. His microscopes were restricted in their resolution, or clarity, which appeared to limit what microbes he was able to observe.

Antony van Leeuwenhoek (1632-1723)

Antony van Leeuwenhoek was a Dutch cloth merchant, who also happened to dabble in microscopes. He constructed a simple microscope (which has a single lens), where the lens was held between two silver plates. Apparently he relished viewing microbes from many different sample types – pond water, fecal material, teeth scrapings, etc. He made detailed drawings and notes about his observations and discoveries, sending them off to the Royal Society of London, the scientific organization of that time. This invaluable record clearly indicates that he saw both bacteria and a wide variety of protists. Some microbiologists refer to van Leeuwenhoek as the "Father of Microbiology," because of his contributions to the field.

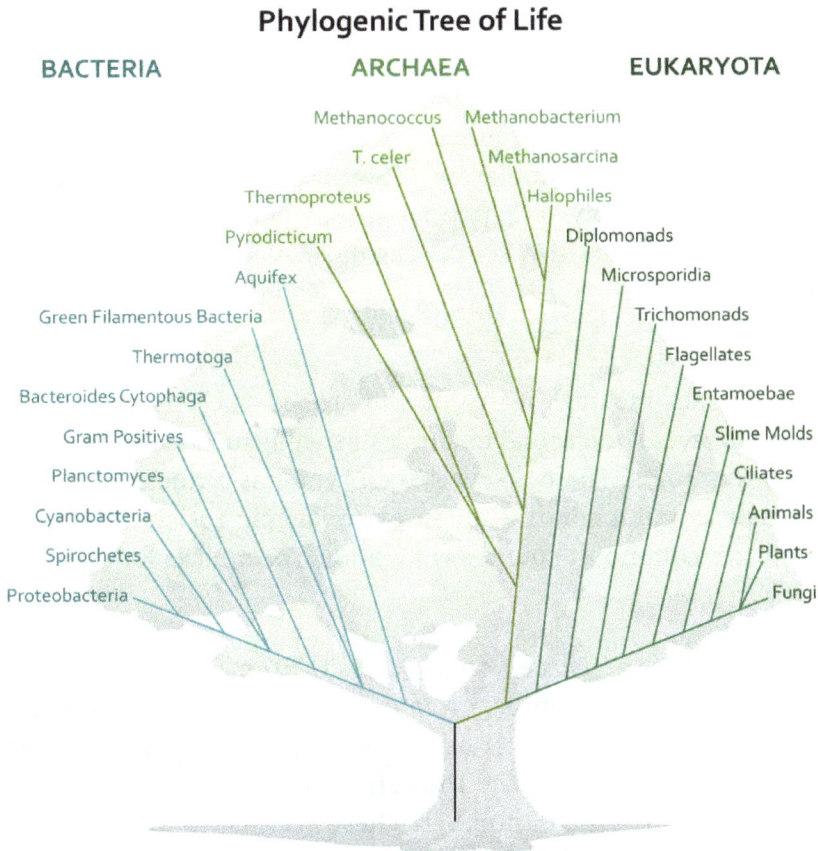

Phylogenic Tree of Life

BACTERIA   ARCHAEA   EUKARYOTA

Methanococcus   Methanobacterium

T. celer   Methanosarcina

Thermoproteus   Halophiles

Pyrodicticum   Diplomonads

Aquifex   Microsporidia

Green Filamentous Bacteria   Trichomonads

Thermotoga   Flagellates

Bacteroides Cytophaga   Entamoebae

Gram Positives   Slime Molds

Planctomyces   Ciliates

Cyanobacteria   Animals

Spirochetes   Plants

Proteobacteria   Fungi

Tree of Life

MICROBIAL GROUPS

Classification of organisms, or the determination of how to group them, continually changes as we acquire new information and new tools of assessing the characteristics of an organism. Currently all organisms are grouped into one of three categories or domains: Bacteria, Archaea, and Eukarya. The Three Domain Classification, first proposed by Carl Woese in the 1970s, is based on ribosomal RNA (rRNA) sequences and widely accepted by scientists today as the most accurate current portrayal of organism relatedness.

The Bacteria domain contains some of the best known microbial examples (E. coli, anyone?). Most of the members are unicellular, cells lack a nucleus or any other organelle, most members have a cell wall with a particular substance known as peptidoglycan (not found anywhere else but in bacteria!), they have 70S ribosomes, and humans are intimately familiar with many members, since they are common in soil, water, our foods, and our own bodies. All Bacteria are considered microbes.

BACTERIA

Archaea is a relatively new domain, since these organisms used to be grouped with the bacteria. There are some obvious similarities, since they are mostly unicellular, cells lack a nucleus or any other organelle, they have 70S ribosomes, and all Archaea are microbes. But they have completely different cell walls that can vary markedly in composition (but notably lack peptidoglycan and might have pseudomurien instead). In addition, their rRNA sequences have shown that they are not closely related to Bacteria at all.

Eukarya

The Eukarya Domain includes many non-microbes, such as animals and plants, but there are numerous microbial examples as well, such as fungi, protists, slime molds, and water molds. The eukaryotic cell type has a nucleus, as well as many organelles, such as mitochondria or an endoplasmic reticulum. They have 80S ribosomes and are commonly found as unicellular or multicellular.

VIRUSES

Viruses are not part of the Three Domain Classification, since they lack ribosomes and therefore lack rRNA sequences for comparison. They are classified separately, using characteristics specific to viruses. Viruses are typically described as "obligate intracellular parasites," a reference to their strict requirement for a host cell in order to replicate or increase in number. These acellular entities are often agents of disease, a result of their cell invasion.

TAXONOMIC RANKS

Taxonomic ranks are a way for scientists to organize information about organisms, by determining relatedness. Domains are the largest grouping used, followed by numerous smaller groupings, where each smaller grouping consists of organisms that share specific features in common. Each level becomes more and more restrictive as to whom can be a member. Eventually we get down to genus and species, the groupings used for formation of a scientific name. This is the binomial nomenclature devised by Carl Linnaeus in the 1750s.

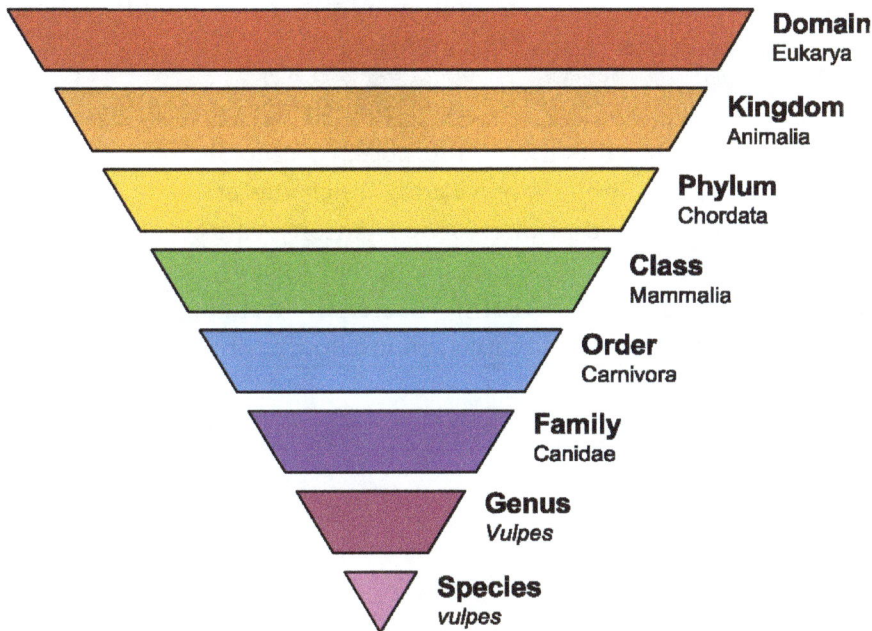

Domain
Eukarya

Kingdom
Animalia

Phylum
Chordata

Class
Mammalia

Order
Carnivora

Family
Canidae

Genus
Vulpes

Species
vulpes

Taxonomic Ranks

BINOMIAL NOMENCLATURE

When referring to the actual scientific name assigned to an organism, it is important to follow convention, so it is clear to everyone that you are referring to the scientific name. There are rules in science (just like in English class, where you would never refer to "mr. robert louis stevenson," or at least not without expecting to get your paper back with red all over it).

A scientific name is composed of a genus and a species, where the genus is a generic name and the species is specific. The species name, once assigned, is permanent for the organism, while the genus can change if new information becomes available. For example, the bacterium previously known as Streptococcus faecalis is now Enterococcus faecalis because sequencing information indicates that it is more closely related to the members of the Enterococcus genus. It is important to note that it is inappropriate to refer to an organism by the species alone (i.e. you should never refer to E. coli as "coli" alone. Other bacteria can have the species "coli" as well.)

Now for the rules: The genus is always capitalized. The species is always lowercase. And both the genus and the species are italicized (common if typewritten) or underlined (common if handwritten). The genus may be shortened to its starting letter, but only if the name has been referred to in the text in its entirety at least once first (the exception to this is E. coli, due to its commonality, where hardly anyone spells out the Escherichia genus anymore).

KEY WORDS

Microbiology, microorganisms, microbes, unicellular, multicellular, differentiation, sterilization, observation, micrometers (μm), nanometers (nm), Robert Hooke, compound microscope, Antony van Leeuwenhoek, simple microscope, Royal Society of London, Father of Microbiology, Three Domain Classification, ribosomal RNA (rRNA), Bacteria, Archaea, Eukarya, obligate intracellular parasites, taxonomic ranks, genus, species, binomial nomenclature.

02 | Introduction to Microscopes

INTRODUCTION: AN OVERVIEW

You have probably figured out that microbes (AKA microorganisms) are pretty small, right? Yeah, well, size isn't everything. But numbers, that is something. If you were to take one gram of soil and start counting the microbes in it at a rate of 1 microbe/second, it would take you over 33 years to complete your counting. Then most of you would be in your 50s and having a mid-life crisis, so let's not go there...but the small size of microbes certainly has made it difficult to study them, particularly in the beginning. (If you want a visual idea of scale, check out the Cell Size and Scale tool, which allows you to zoom in from a coffee bean to a carbon atom. Be sure to pay careful attention to the microbes in between!)

O.k., if you want to see something really, really, really small, who ya gonna call? Not Ghostbusters™, that's for sure. I would try someone with a microscope. (Microscope Man? Maybe not.) Now I will admit, with the advent of molecular biology there's a lot of microbiology nowadays that happens without a microscope. But if you want to actually visualize microbes, you'll need the ability to magnify – you'll need a microscope of some type. And, since "seeing is believing," it was the visualization of microbes that got people interested in them in the first place.

MICROSCOPY IN THE 1600S

It is believed that Robert Hooke was one of the first scientists to actually observe microbes, in 1665. His illustrations and observations from a variety of objects viewed under a microscope were published in the book Micrographia. Hooke used a compound microscope, meaning it contained two sets of lenses for magnification: the ocular lens next to the eye and the objective lens, next to the specimen or object. The magnification of a compound microscope is a product of the ocular lens magnification and the objective lens magnification. Thus a microscope with an ocular magnification of 10x and an objective magnification of 50x would have a total magnification of 500x. You can see a drawing of Hooke's microscope.

Antonie van Leeuwenhoek, often called the "Father of Microbiology," wasn't a scientist by profession. He was a cloth merchant from Holland who was believed to be inspired by Mr. Hooke's work, probably with the original intention of examining textiles to determine quality. Very quickly van Leeuwenhoek started examining just about everything under the microscope and we know this because he kept detailed notes about both his samples and his observations. Van Leeuwenhoek was using what is called a simple microscope, a microscope with just a single lens. Essentially, it is a magnifying glass. But the lenses that he produced were of such high quality that he is given credit for the discovery of single-celled life forms. You can learn more about van Leeuwenhoek's observations.

MODERN MICROSCOPY: LIGHT MICROSCOPES

Let us face it, a modern microscope is a pretty technical tool, even one of the cheaper versions. If you want to understand the limitations of a light microscope you have to understand concepts like resolution, wavelength, and numerical aperture, where their relationship to one another is summed up by the Abbé equation:

$$d = \frac{0.5\,\lambda}{n\,\sin\theta}$$

λ = wavelength of light
d = minimal distance to distinguish between two close objects
$n\sin\theta$ = numerical aperture

Abble Equation

In microscopy the definition of resolution is typically the ability of a lens to distinguish two objects that are close together. So, in the Abbé equation d becomes the minimal distance where two objects next to one another can be resolved or distinguished as individual objects. Resolution is dependent upon the wavelength of illumination being used, where a shorter wavelength will result in a smaller d. Lastly, we have the effect of the numerical aperture, which is a function of the objective lens and its ability to gather light. The numerical aperture value is actually defined by two components: n, which is the refractive index of the medium the lens is working in, and sin Θ, which is a measurement of the cone of light that enters the objective. A lens can typically work in two media: air, with a refractive index of 1.00, or oil, with a refractive index of 1.25. Oil will allow more of the light to be collected, by directing more of the light rays into the objective lens. The maximum total magnification for a microscope using visual light for illumination is around 1500X, where the microscope might have 15x oculars and a 100x oil immersion objective. The highest resolution possible is around 0.2 μm. If objects or cells are closer together than this, they can't be distinguished as separate entities.

Here's a nice description from Nikon, including an interactive tutorial on numerical aperture and image resolution. And then there are so many microscopes, so little time! The type that you need depends upon the specific type of microbes that you want to visualize.

For light microscopes there are six different types of microscopes, all using light as the source of illumination: bright-field microscope, dark-field microscope, phase contrast microscope, differential interference contrast (DIC), fluorescence microscope, and confocal scanning laser microscope (CSLM). Let us look at the details of each type:

BRIGHT-FIELD MICROSCOPE

The bright-field microscope is your standard microscope that you could purchase for your niece or nephew at any toy store. Here is a website on the basic parts of a bright-field compound microscope, in case you are not enrolled in the general microbiology lab. The specimen is illuminated by a light source at the base of the microscope and then initially magnified by the objective lens, before being magnified again by the ocular lens. Remember that the total magnification achieved is a product of the magnification of both lenses.

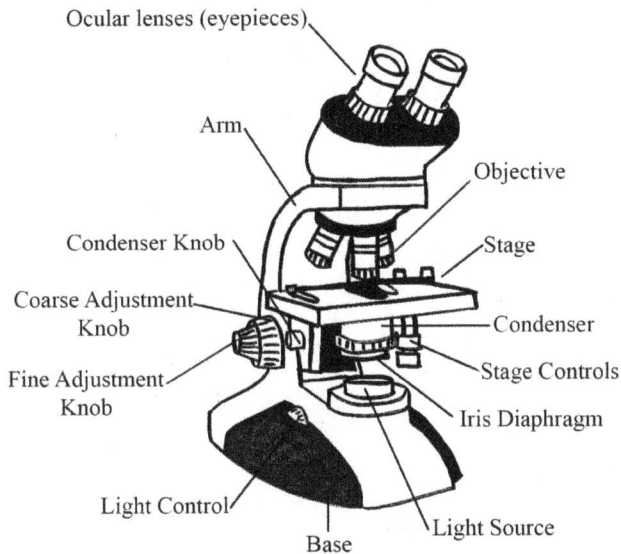

Ocular lenses (eyepieces)
Arm
Objective
Condenser Knob
Stage
Coarse Adjustment Knob
Condenser
Fine Adjustment Knob
Stage Controls
Iris Diaphragm
Light Control
Light Source
Base

Bright-Field Microscope

The specimen is typically visualized because of differences in contrast between the specimen itself and its surrounding environment. But that does not apply to unstained bacteria, which have very little contrast with their environment, unless the cells are naturally pigmented. That is why staining (see section below) is such an important concept in microscopy. A bright-field microscope will work reasonably well to view the larger eukaryotic microbes (i.e. protozoa, algae) without stain, but unstained bacteria will be almost invisible. Stained bacteria will appear dark against a bright background (ah, I knew that there was a reason for the term "bright-field.")

DARK-FIELD MICROSCOPE

The dark-field microscope is really just a slightly modified bright-field microscope. In fact, you could make this modification to the microscope you have at home! It makes use of what is known as a dark-field stop, an opaque disk that blocks light directly underneath the specimen so that light reaches it from the sides. The result is that only light that has been reflected or refracted by the specimen will be collected by the objective lens, resulting in cells that appear bright against a dark background (thus the term "dark-field." Yes, it's all making sense now). This allows for observation of living, unstained cells which is particularly nice if you want to observe motility or eukaryotic organelles.

PHASE CONTRAST MICROSCOPE

The phase-contrast microscope is also a modified bright-field microscope, although the modifications are getting more complex, as well as more expensive. This microscope also uses an opaque ring or annular stop, but this one has a transparent ring that only releases light in a hollow cone. The principle of this microscope gets back to the idea of refractive index and the fact that cells have a different refractive index than their surroundings, resulting in light that differs slightly in phase. The difference is amplified by a phase ring found in a special phase objective. The phase differences can be translated into differences in brightness, resulting in a dark image amidst a bright background. This allows for the observation of living, unstained cells, once again useful to observe motility or eukaryotic organelles.

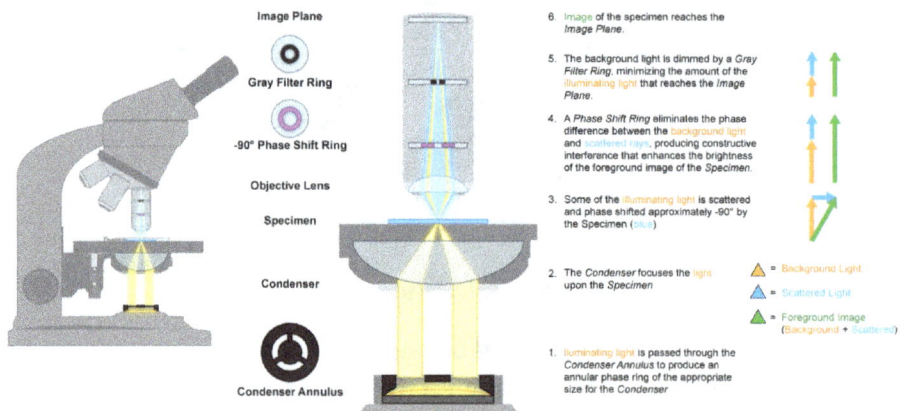

Phase Contrast Microscope Mechanics.

Differential Interference Contrast (DIC) Microscope

The differential interference contrast microscope operates on much the same principle as the phase-contrast microscope, by taking advantage of the differences in refractive index of a specimen and its surroundings. But it uses polarized light that is then split into two beams by a prism. One beam of light passes through the specimen, the other passes through the surrounding area. When the beams are combined via a second prism they "interfere" with one another, due to being out of phase. The resulting images have an almost 3D effect, useful for observing living, unstained cells.

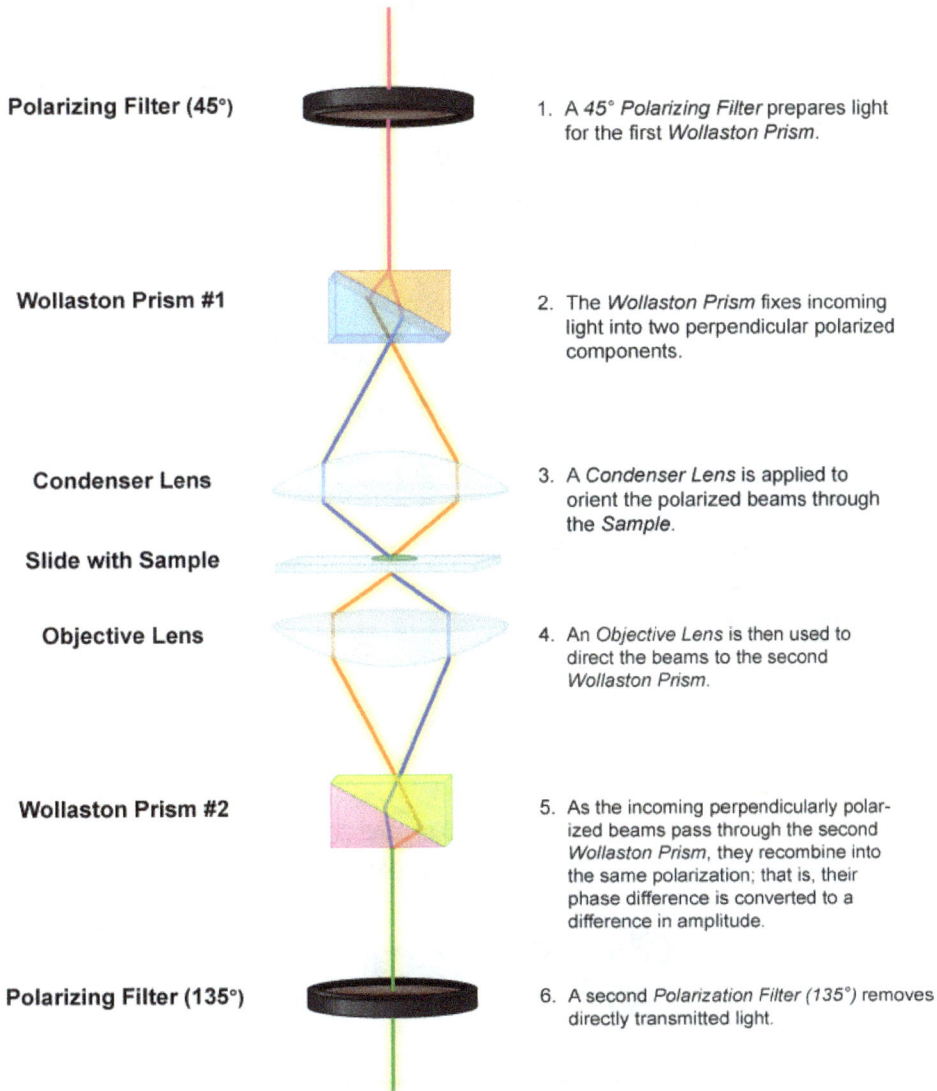

Polarizing Filter (45°)	1. A *45° Polarizing Filter* prepares light for the first *Wollaston Prism*.
Wollaston Prism #1	2. The *Wollaston Prism* fixes incoming light into two perpendicular polarized components.
Condenser Lens	3. A *Condenser Lens* is applied to orient the polarized beams through the *Sample*.
Slide with Sample	
Objective Lens	4. An *Objective Lens* is then used to direct the beams to the second *Wollaston Prism*.
Wollaston Prism #2	5. As the incoming perpendicularly polarized beams pass through the second *Wollaston Prism*, they recombine into the same polarization; that is, their phase difference is converted to a difference in amplitude.
Polarizing Filter (135°)	6. A second *Polarization Filter (135°)* removes directly transmitted light.

Differential Contrast Microscope Mechanism

FLUORESCENCE MICROSCOPE

A fluorescence microscope utilizes light that has been emitted from a specimen, rather than passing through it. A mercury-arc lamp is used to generate an intense beam of light that is filtered to produce a specific wavelength of light directed at the specimen by use of dichromatic mirror, which reflects short wavelengths and transmits longer wavelengths. Naturally fluorescent organisms will absorb the short-wavelengths and emit fluorescent light with a longer wavelength that will pass through the dichromatic mirror and can be visualized. There are a variety of microbes with natural fluorescence but there are certainly far more organisms that lack this quality. Visualization of the latter organisms depends upon the use of fluorochromes, fluorescent dyes that bind to specific cell components. The fluorochromes can also be attached to antibodies, to highlight specific structures or areas of the cell, or even different organisms.

Fluorescence Microscope. By Masur (Own work) [GFDL, CC-BY-SA-3.0 or CC BY-SA 1.0-2.0-2.5], via Wikimedia Commons

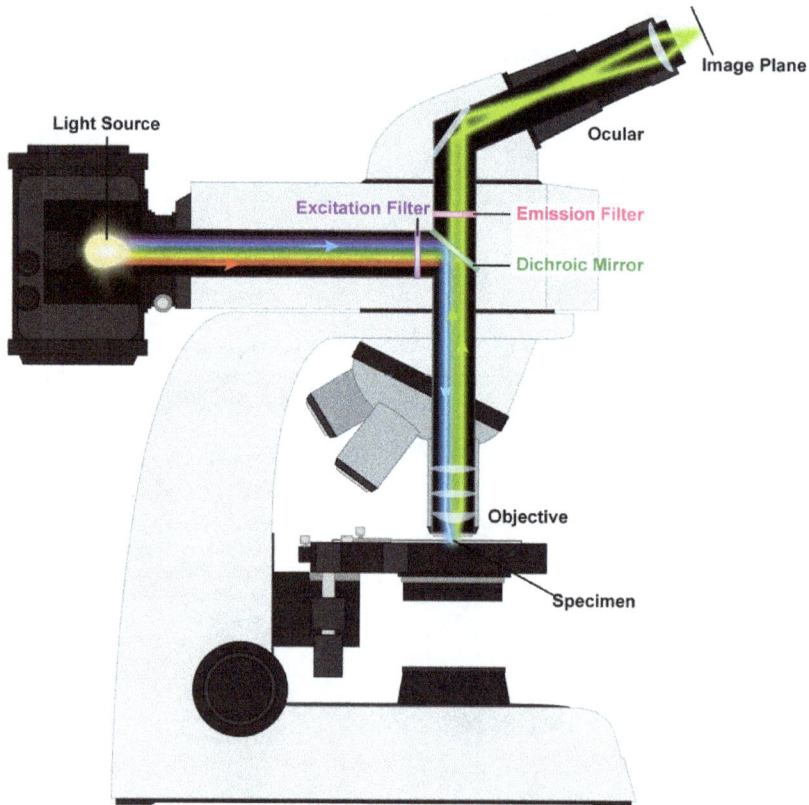

Fluorescence Microscope Mechanism

CONFOCAL SCANNING LASER MICROSCOPE (CSLM)

In order to understand how a confocal scanning laser microscope works, it is helpful to understand how a fluorescence microscope works, so hopefully you already read the previous section. A CSLM uses a laser for illumination, due to the high intensity. The light is directed at dichromatic mirrors that move, "scanning" the specimen. The longer wavelengths emitted by the fluorescently stained specimen pass back through the mirrors, through a pinhole, and are measured by a detector. The pinhole serves to conjugate the focal point of the lens (ah, that's where the term confocal came from!), which means it allows for complete focus of a given point. Since the entire specimen is scanned in the x-z planes (all three axes), the information acquired by the detector can be compiled by a computer to create a single 3D image entirely in focus. This is a particularly useful tool for viewing complex structures such as biofilms.

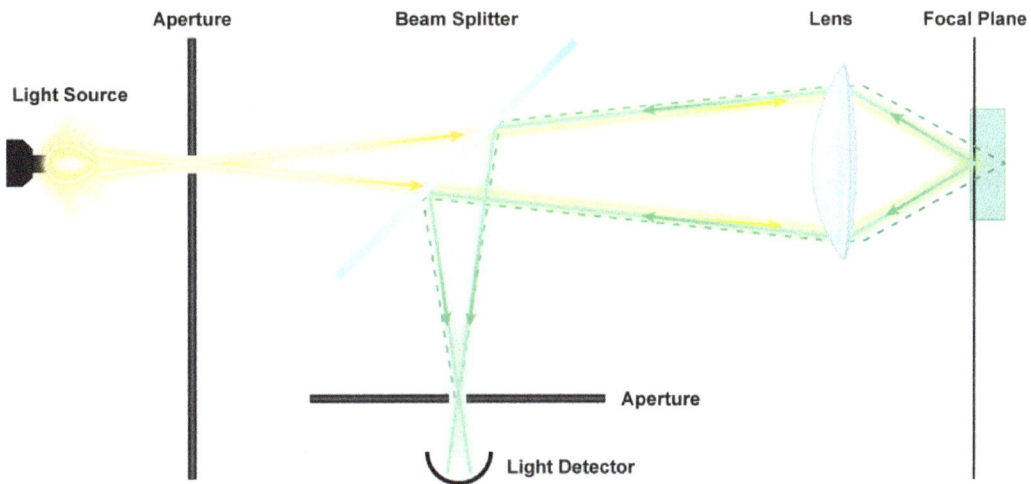

Confocal Scanning Laser Microscope Mechanism

STAINING

Most of the microbes, particularly unicellular microbes, would not be apparent without the help of staining. It helps to make something so small a bit easier to see, by providing contrast between the microorganism and its background. A simple stain makes use of a single dye, either to stain the cells directly (direct stain) or to stain the background surrounding the cells (negative stain). From this a researcher can gather basic information about a cell's size, morphology (shape), and cell arrangement.

There are also more complex stains, known as differential stains, that combine stains to allow for differentiation of organisms based on their characteristics. The Gram stain, developed in 1884, is the most common differential stain used in microbiology, where bacterial cells are separated based on their cell wall type: gram positive bacteria which stain purple and gram negative bacteria which stain pink. Some bacteria have a specialized cell wall that must be stained with the acid-fast stain, where acid-fast bacteria stain red and non-acid-fast bacteria stain blue. Other differential stains target specific bacterial structures, such as endospores, capsules, and flagella, to be talked about later.

EVEN MORE MODERN MICROSCOPY: ELECTRON MICROSCOPES

Light microscopes are great if you are observing eukaryotic microbes and they might work for observing bacteria and archaea, but they are not going to work at all to observe viruses. Remember that the limit of resolution for a light microscope is 0.2 μm or 200 nm and most viruses are smaller than that. So, we need something more powerful. Enter the electron microscopes, which replace light with electrons for visualization. Since electrons have a wavelength of 1.23 nm (as opposed to the 530 nm wavelength of blue-green light), resolution increases to around 0.5 nm, with magnifications over 150,000x. The drawback of using electrons is that they must be contained in a vacuum, eliminating the possibility of working with live cells. There is also some concern that the extensive sample preparation might distort the specimen's characteristics or cause artifacts to form.

There are two different types of electron microscope, the transmission electron microscope (TEM) and the scanning electron microscope (SEM):

Transmission Electron Microscope (TEM)

The transmission electron microscope utilizes an electron beam directed at the specimen with the use of electromagnets. Dense areas scatter the electrons, resulting in a dark area on the image, while electrons can pass (or "transmit") through the less dense areas, resulting in a brighter section. The image is generated on a fluorescent screen and can then be captured.

Since the electrons are easily scattered by extremely thick specimens, samples must be sliced down to 20-100 nm in thickness, typically by being embedded in some type of plastic and then being cut with a diamond knife into extremely thin sections. The resulting pictures represent one slice or plane of the specimen.

Scanning Electron Microscope (SEM)

The scanning electron microscope also utilizes an electron beam but the image is formed from secondary electrons that were released from the surface of the specimen and then collected by a detector. More electrons are released from raised areas of the specimen, while less secondary electrons will be collected from sunken areas. In addition, the electron beam is scanned over the specimen surface, producing a 3D image of the external features.

- Electron Gun
- Heated Filaments
- Anode
- Gun Alignment Coils
- Condenser Aperture
- Condenser Lens
- Secondary Condenser Aperture
- Objective Stigmator
- Sample Holder
- Objective Lens
- Diffraction Lens
- Objective Aperture
- Intermediate Lens
- Diffraction Stigmator
- Projector Lens 1
- Projector Lens 2
- Flourescent Viewing Screen

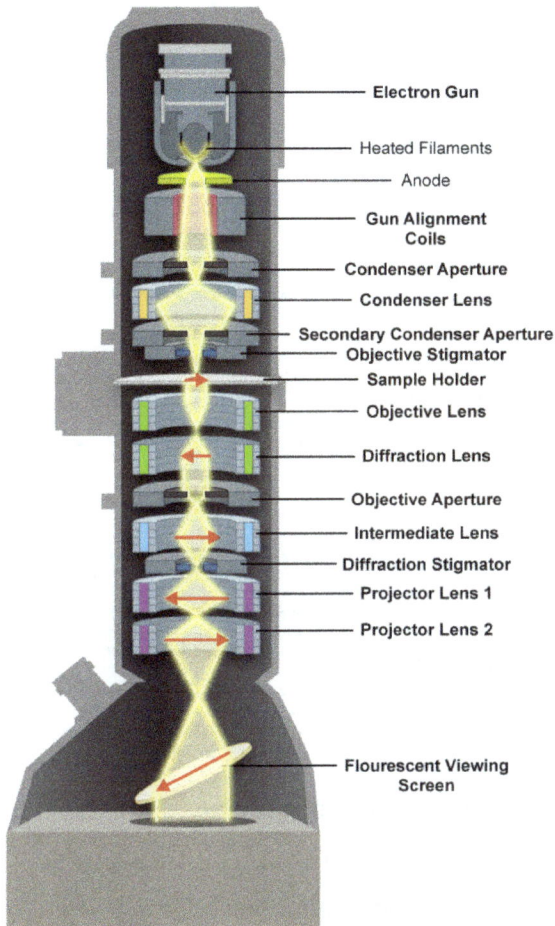

1. Electrons are emitted into a vacuum tube by heating *Cathode Filaments* in the electron gun
2. The cathode ray then passes through an *Anode*, which accelerates and focues the beam; *Alignment Coils* additionally accelerate the beam
3. An adjustable *Condenser Aperture* prepares the beam for the *Condenser Lens* by blocking off-axis or off-energy electrons from proceeding.
4. The magnetic *Condenser Lens* applies a magnetic field, inducing a helical path for the electrons, and leading the cone-shaped electron beam to converge on a spot
5. A *Stigmator* helps to adjust the beam and prevent astigmatism (different foci between rays) in the optical system
6. Electrons pass through the thinly sliced sample, inserted onto a grid-like stage
7. The *Objective Lens* focuses the image of the sample
8. A *Diffraction Lens* is used to apply Bragg Scattering to the electrons
9. The *Objective Aperture*, positioned on the back focal plane of the scattered rays, selects (or excludes) the portion of the sample that produced the scattering
10. *Projector Lenses* calibrate the magnification of the image
11. The image is visualized through oculars or by an image recording system underneath the *Fluorescent Screen*

Transmission Electron Microscope Transmission

If you want to see some beautiful TEM and SEM photomicrographs, check out Dennis Kunkel's site. Most have been colorized, but they are quite stunning. On the other end of the spectrum, here are pictures taken with the Intel Play QX3, a plastic microscope for children. Be careful, you could get lost in this website. But it is great to see what an inexpensive microscope can produce in the hands of someone who knows what they're doing! These pictures are stunning as well.

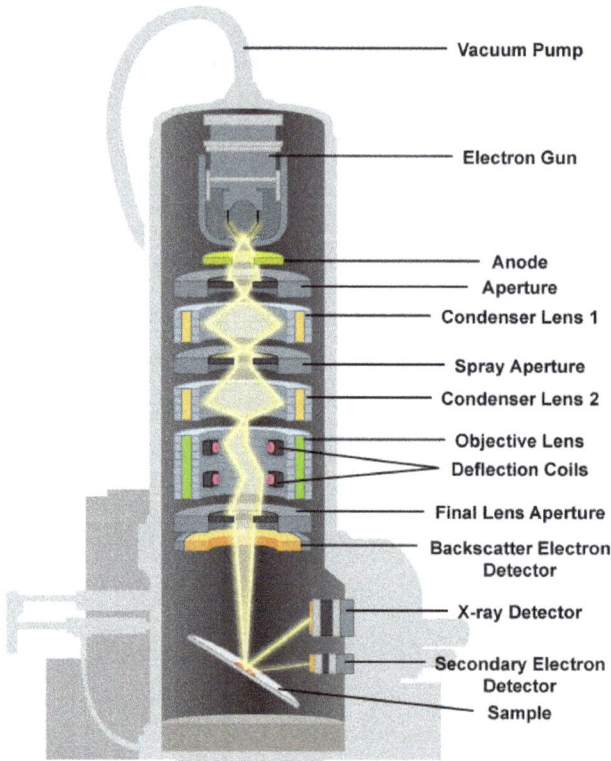

Vacuum Pump

Electron Gun

Anode
Aperture
Condenser Lens 1

Spray Aperture

Condenser Lens 2

Objective Lens
Deflection Coils

Final Lens Aperture

Backscatter Electron
Detector

X-ray Detector

Secondary Electron
Detector
Sample

1. Electrons are emitted into a vacuum tube by heating cathode filaments in the *Electron Gun*

2. The cathode ray then passes through an *Anode*, which accelerates and focues the beam

3. The *Condenser Aperture* prepares the beam for the *Condenser Lens* by blocking off-axis or off-energy electrons from proceeding

4. The *Spray Aperture* works in conjunction with magnetic *Condenser Lenses,* which apply a magnetic field to the beam, inducing a helical path that focuses the beam onto a spot

5. Within the *Objective Lens* is embedded pairs of *Deflection Coils* which deflect the electron beam to produce a rasterized scan of the sample

6. Electrons in the beam interact with the sample. As they do so, they will randomly scatter and absorb within the sample.

7. The X-ray Detector is the primary detector of the high-energy electrons, which can map the sample.

The Secondary Electron detector picks up electrons scattered from the sample surface, generating a topographic image.

The Backscatter Electron Detector identifies chemical phase differences in the sample by picking up electrons scattered from the interaction volume of the specimen.

Scanning Electron Microscope Mechanism.

Scanning Electron Microscope

WELCOME TO 21ST CENTURY MICROSCOPY: SCANNING PROBE MICROSCOPES

As technology has advanced, even more powerful microscopes have been invented, ones that can even allow for visualization at the atomic level. These microscopes can be used in microbiology but more often they are used in other fields, to allow visualization of chemicals, metals, magnetic samples, and nanoparticles, wherever the 0.1 nm resolution and 100,000,000x magnification might be needed.

The scanning probe microscopes are thus named because they move some type of probe over a specimen's surface in the x-z planes, allowing computers to generate an extremely detailed 3D image of the specimen. Resolution is so high because the probe size is much smaller than the wavelength of either visible light or electrons. Both microscopes can be used to study objects in liquid, allowing for the examination of biological molecules. There are two different types of scanning probe microscopes, the scanning tunneling microscope (STM) and the atomic force microscope (AFM):

SCANNING TUNNELING MICROSCOPE (STM)

The scanning tunneling microscope has an extremely sharp probe, 1 atom thick, that maintains a constant voltage with the specimen surface allowing electrons to travel between them. This tunneling current is maintained by raising and lowering the probe to sustain a constant height above the sample. The resulting motion is tracked by a computer, which generates the final image.

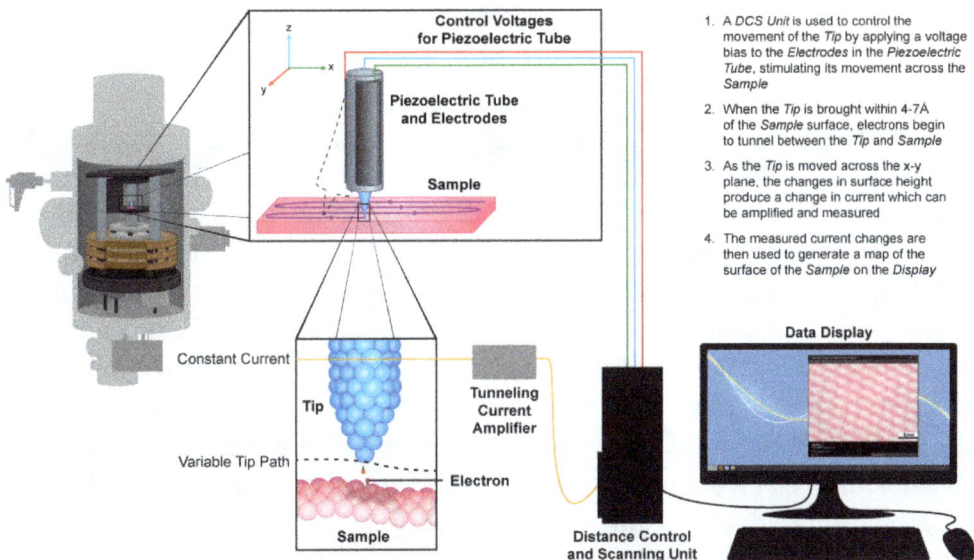

1. A *DCS Unit* is used to control the movement of the *Tip* by applying a voltage bias to the *Electrodes* in the *Piezoelectric Tube*, stimulating its movement across the *Sample*

2. When the *Tip* is brought within 4-7Å of the *Sample* surface, electrons begin to tunnel between the *Tip* and *Sample*

3. As the *Tip* is moved across the x-y plane, the changes in surface height produce a change in current which can be amplified and measured

4. The measured current changes are then used to generate a map of the surface of the *Sample* on the *Display*

Scanning Tunneling Microscope Mechanism.

ATOMIC FORCE MICROSCOPE (AFM)

The atomic force microscope was developed as an alternative to the STM, for use with samples that do not conduct electricity well. The microscope utilizes a cantilever with an extremely sharp probe tip that maintains a constant height above the specimen, typically by direct contact with the sample. Movement of the cantilever to maintain this contact deflects a laser beam, translating into an image of the object. Once again, computers are used to generate the image.

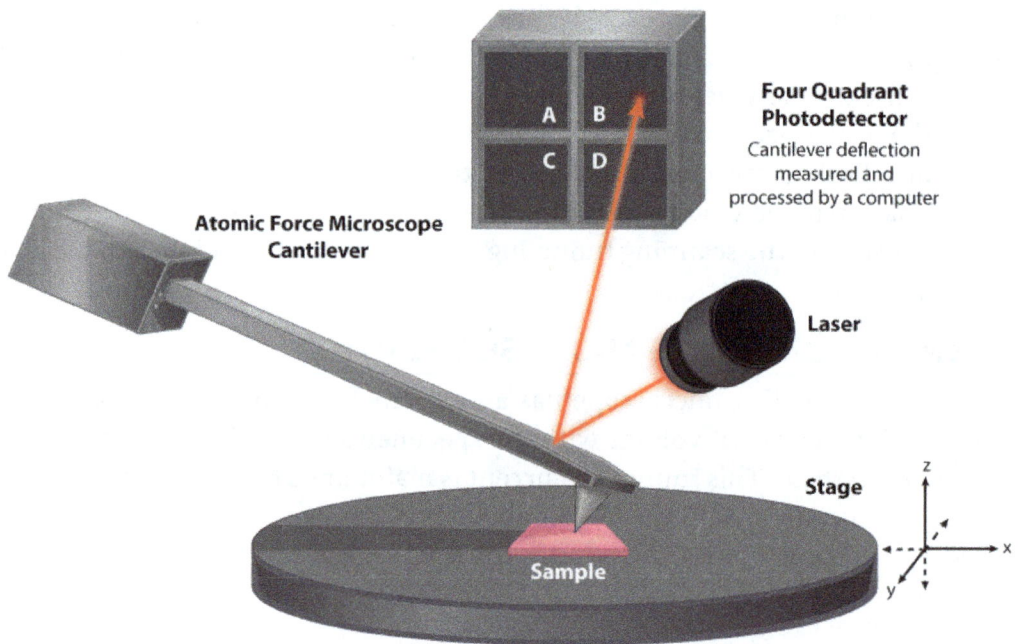

Atomic Force Microscope Mechanism.

KEY WORDS

Magnification, Robert Hooke, compound microscope, ocular lens, objective lens, total magnification, van Leeuwenhoek, simple microscope, Abbé equation, resolution, wavelength, numerical aperture, refractive index, oil immersion objective, light microscope, bright-field microscope, dark-field microscope, phase contrast microscope, differential interference contrast (DIC) microscope, fluorescence microscope, fluorochrome, confocal scanning laser microscope (CSLM), simple stain, direct stain, negative stain, differential stain, Gram stain, acid-fast stain, electron microscope (EM), transmission electron microscope (TEM), scanning electron microscope (SEM), scanning probe microscopes, scanning tunneling microscopes (STM), atomic force microscope (AFM).

INTRODUCTION

Traditionally, cellular organisms have been divided into two broad categories, based on their cell type. They are either prokaryotic or eukaryotic. In general, prokaryotes are smaller, simpler, with a lot less stuff, which would make eukaryotes larger, more complex, and more cluttered. The crux of their key difference can be deduced from their names: "karyose" is a Greek word meaning "nut" or "center," a reference to the nucleus of a cell. "Pro" means "before," while "Eu" means "true," indicating that prokaryotes lack a nucleus ("before a nucleus") while eukaryotes have a true nucleus. More recently, microbiologists have been rebelling against the term prokaryote because it lumps both bacteria and the more recently discovered archaea in the same category. Both cells are prokaryotic because they lack a nucleus and other organelles (such as mitochondria, Golgi apparatus, endoplasmic reticulum, etc), but they aren't closely related genetically. So, to honor these differences this text will refer to the groups as the archaea, the bacteria, and the eukaryotes and try to leave the prokaryotic reference out of it.

CELL MORPHOLOGY

Cell morphology is a reference to the shape of a cell. It might seem like a trivial concept but to a cell it is not. The shape dictates how that cell will grow, reproduce, obtain nutrients, move, and it's important to the cell to maintain that shape to function properly. Cell morphology can be used as a characteristic to assist in identifying particular microbes but it's important to note that cells with the same morphology are not necessarily related.

Bacteria tend to display the most representative cell morphologies, with the most common examples listed here:

⊙ Coccus (pl. cocci) – a coccus is a spherically shaped cell.

⊙ Bacillus (pl. bacilli) – a bacillus is a rod-shaped cell.

Coccus   Bacillus   Curved Rods   Pleomorphic

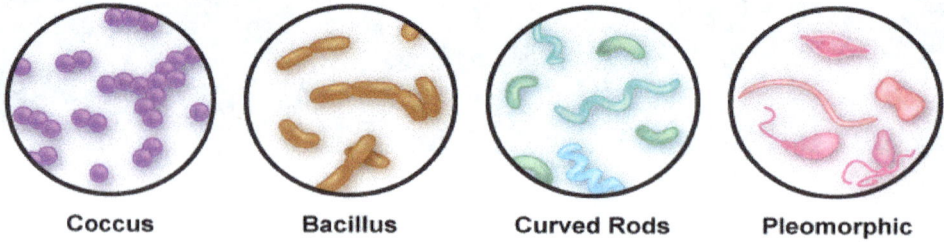

Bacterial Cell Morphologies.

- Curved rods – obviously this is a rod with some type of curvature. There are three sub-categories: the vibrio, which are rods with a single curve and the spirilla/spirochetes, which are rods that form spiral shapes. Spirilla and spirochetes are differentiated by the type of motility that they exhibit, which means it is hard to separate them unless you are looking at a wet mount.

- Pleomorphic – pleomorphic organisms exhibit variability in their shape.

There are additional shapes seen for bacteria, and an even wider array for the archaea, which have even been found as star or square shapes. Eukaryotic microbes also tend to exhibit a wide array of shapes, particularly the ones that lack a cell wall such as the protozoa.

CELL SIZE

Cell size, just like cell morphology, is not a trivial matter either, to a cell. There are reasons why most archaeal/bacterial cells are much smaller than eukaryotic cells. Much of it has to do with the advantages derived from being small. These advantages relate back to the surface-to-volume ratio of the cell, a ratio of the external cellular layer in contact with the environment compared to the liquid inside. This ratio changes as a cell increases in size. Let us look at a 2 µm cell in comparison with a cell that is twice as large at 4 µm.

r = 1 µm
surface area = 12.6 µm2
volume = 4.2 µm3

r = 2 µm
surface area = 50.3 µm2
volume = 33.5 µm3

The surface-to-volume ratio of the smaller cell is 3, while the surface-to-volume ratio of the larger cell decreases to 1.5. Think of the cell surface as the ability of the cell to bring in nutrients and let out waste products. The larger the surface area, the more possibilities exist for engaging in these activities. Based on this, the larger cell would have an advantage. Now think of the volume as representing what the cell has to support. As the surface-to-volume ratio goes down, it indicates that the cell has less of an opportunity to bring in the nutrients needed to support the cell's activities – activities such as growth and reproduction. So, small cells grow and reproduce faster. This also means that they evolve faster over time, giving them more opportunities to adapt to environments.

Keep in mind that the size difference (bacterial/archaeal cells = smaller, eukaryotic cells = larger) is on average. A typical bacterial/archaeal cell is a few micrometers in size, while a typical eukaryotic cell is about 10x larger. There are a few monster bacteria that fall outside the norm in size and still manage to grow and reproduce very quickly. One such example is Thiomargarita namibiensis, which can measure from 100-750 µm in length, compared to the more typical 4 µm length of E. coli. T. namibiensis manages to maintain its rapid reproductive rate by producing very large vacuoles or bubbles that occupy a large portion of the cell. These vacuoles reduce the volume of the cell, increasing the surface-to-volume ratio. Other very large bacteria utilize a ruffled membrane for their outer surface layer. This increases the surface area, which also increase the surface-to-volume ratio, allowing the cell to maintain its rapid reproductive rate.

CELL COMPONENTS

All cells (bacterial, archaeal, eukaryotic) share four common components:

- ◉ **Cytoplasm** – cytoplasm is the gel-like fluid that fills each cell, providing an aqueous environment for the chemical reactions that take place in a cell. It is composed of mostly water, with some salts and proteins.

- ◉ **DNA** – deoxyribonucleic acid or DNA is the genetic material of the cell, the instructions for the cell's abilities and characteristics. This complete set of genes, referred to as a genome, is localized in an irregularly-shaped region known as the nucleoid in bacterial and archaeal cells, and enclosed into a membrane-bound nucleus in eukaryotic cells.

- ◉ **Ribosomes** – the protein-making factories of the cell are the ribosomes. Composed of both RNA and protein, there are some distinct differences between the ones found in bacteria/archaea and the ones found in eukaryotes, particularly in terms of size and location. The ribosomes of bacteria and archaea are found floating in the cytoplasm, while many of the eukaryotic ribosomes are organized along the endoplasmic reticulum, a eukaryotic organelle. Ribosomes are measured using the Svedberg unit, which corresponds to the rate of sedimentation when centrifuged. Bacterial/ archaeal ribosomes have a measurement of 70S as a sedimentation value, while eukaryotic ribosomes have a measurement of 80S, an indication of both their larger size and mass.

◉ **Plasma Membrane** – one of the outer boundaries of every cell is the plasma membrane or cell membrane. (A plasma membrane can be found elsewhere as well, such as the membrane that bounds the eukaryotic nucleus, while the term cell membrane refers specifically to this boundary of the cell proper). The plasma membrane separates the cell's inner contents from the surrounding environment. While not a strong layer, the plasma membrane participates in several crucial processes for the cell, particularly for bacteria and archaea, which typically only have the one membrane:

◉ **Acts as a semi**-permeable barrier to allow for the entrance and exit of select molecules. It functions to let in nutrients, excrete waste products, and possibly keep out dangerous substances such as toxins or antibiotics.

◉ Performs metabolic processes by participating in the conversion of light or chemical energy into a readily useable form known as ATP. This energy conservation involves the development of a proton motive force (PMF), based on the separation of charges across the membrane, much like a battery.

◉ "Communicates" with the environment by binding or taking in small molecules that act as signals and provide information important to the cell. The information might relate to nutrients or toxin in the area, as well as information about other organisms.

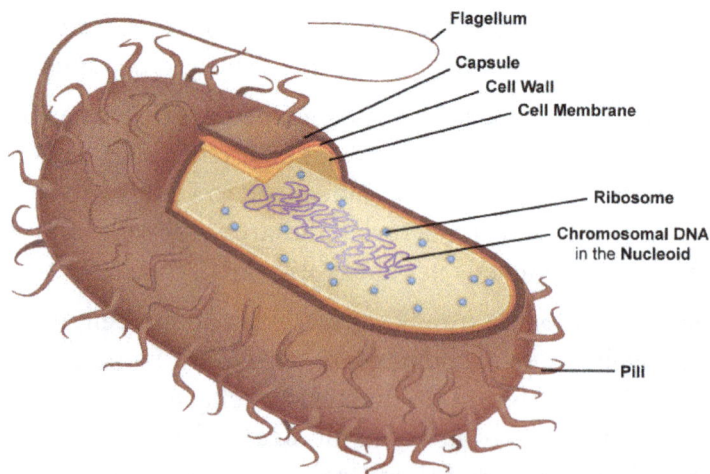

Typical Prokaryotic Cell.

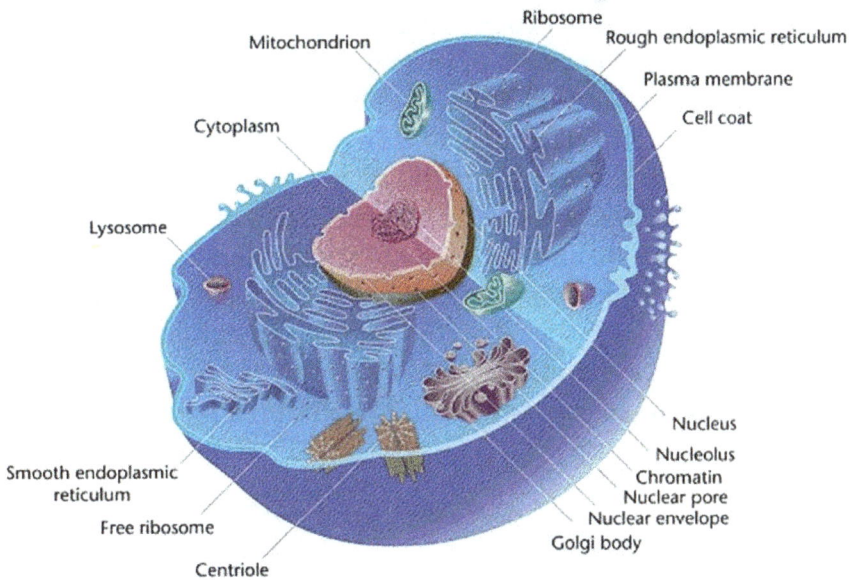

Ribosome
Rough endoplasmic reticulum
Mitochondrion
Plasma membrane
Cytoplasm
Cell coat
Lysosome
Nucleus
Nucleolus
Chromatin
Smooth endoplasmic
reticulum
Nuclear pore
Nuclear envelope
Free ribosome
Golgi body
Centriole

Typical Eukaryotic Cell

Eukaryotes have numerous additional components called organelles, such as the nucleus, the mitochrondria, the endoplasmic reticulum, the Golgi apparatus, etc. These are all membrane-bound compartments that house different activities for the cell. Because each structure is bounded by its very own plasma membrane, it provides the cell with multiple locations for membranous functions to occur.

PLASMA MEMBRANE STRUCTURE

When talking about the details of the plasma membrane it gets a little bit complicated, since bacteria and eukaryotes share the same basic structure, while archaea have marked differences. For now let us cover the basic structure, while the archaeal modifications and variations will be covered in the chapter on archaea.

The plasma membrane is often described by the fluid-mosaic model, which accounts for the movement of various components within the membrane itself. The general structure is explained by the separation of individual substances based on their attraction or repulsion of water. The membrane is typically composed of two layers (a bilayer) of phospholipids, which form the basic structure. Each phospholipid is composed of a polar region that is hydrophilic ("water loving") and a non-polar region that is hydrophobic ("water fearing"). The phospholipids will spontaneously assemble in such a way as to keep the polar regions in contact with the aqueous environment outside of the cell and the cytoplasm inside, while the non-polar regions are sequestered in the middle, much like the jelly in a sandwich.

The phospholipids themselves are composed of a negatively-charged polar head which is a phosphate group, connected by a glycerol linkage to two fatty acid tails. The phosphate group is hydrophilic while the fatty acid tails are hydrophobic. While the membrane is not considered to be particularly strong, it is strengthened somewhat by the presence of additional lipid components, such as the steroids in eukaryotes and the sterol-like hopanoids in bacteria. Embedded and associated with the phospholipid bilayer are various proteins, with myriad functions. Proteins that are embedded within the bilayer itself are called integral proteins while proteins that associate on the outside of the membrane are called peripheral proteins. Some of the peripheral proteins are anchored to the membrane via a lipid tail, and many associate with specific integral proteins to fulfill cellular functions. Integral proteins are the dominant type, representing about 70-80% of the proteins associated with a plasma membrane, while the peripheral proteins represent the remaining 20-30%.

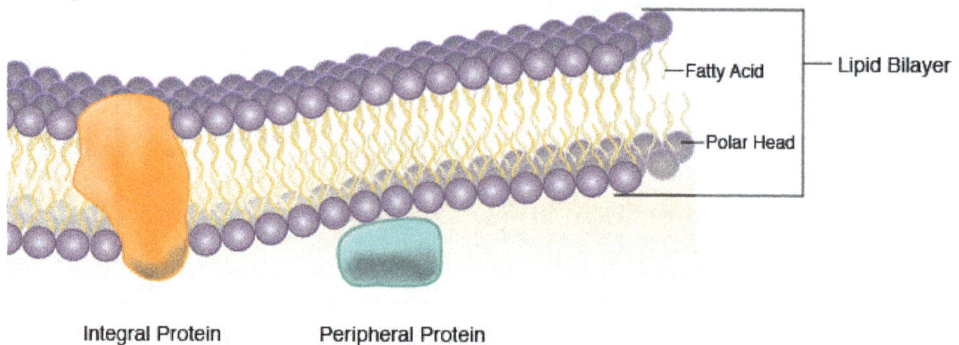

Plasma Membrane Structure.

The amount of protein composing a plasma membrane, in comparison to phospholipid, differs by organism. Bacteria have a very high protein to phospholipid ratio, around 2.5:1, while eukaryotes exhibit a ratio of 1:1, at least in their cell membrane. But remember that eukaryotes have multiple plasma membranes, one for every organelle. The protein to phospholipid ratio for their mitochondrial membrane is 2.5:1, just like the bacterial plasma membrane, providing additional evidence for the idea that eukaryotes evolved from a bacterial ancestor.

KEY WORDS

prokaryote, eukaryote, morphology, coccus, bacillus, vibrio, spirilla, spirochete, pleomorphic, surface-to-volume (S/V) ratio, cytoplasm, DNA, genome, nucleoid, nucleus, ribosome, Svedberg unit, plasma membrane, cell membrane, proton motive force (PMF), fluid-mosaic model, phospholipid, hydrophilic, hydrophobic, polar head, phosphate group, glycerol linkage, fatty acid tail, steroids, hopanoids, phospholipid bilayer, integral protein, peripheral protein.

INTRODUCTION

It is important to note that not all bacteria have a cell wall. Having said that though, it is also important to note that most bacteria (about %90) have a cell wall and they typically have one of two types: a gram positive cell wall or a gram negative cell wall.

The two different cell wall types can be identified in the lab by a differential stain known as the Gram stain. Developed in 1884, it's been in use ever since. Originally, it was not known why the Gram stain allowed for such reliable separation of bacterial into two groups. Once the electron microscope was invented in the 1940s, it was found that the staining difference correlated with differences in the cell walls. Here is a website that shows the actual steps of the Gram stain. After this stain technique is applied the gram positive bacteria will stain purple, while the gram negative bacteria will stain pink.

Gram + Bacteria	Gram - Bacteria

OVERVIEW OF BACTERIAL CELL WALLS

A cell wall, not just of bacteria but for all organisms, is found outside of the cell membrane. It's an additional layer that typically provides some strength that the cell membrane lacks, by having a semi-rigid structure.

Both gram positive and gram negative cell walls contain an ingredient known as peptidoglycan (also known as murein). This particular substance hasn't been found anywhere else on Earth, other than the cell walls of bacteria. But both bacterial cell wall types contain additional ingredients as well, making the bacterial cell wall a complex structure overall, particularly when compared with the cell walls of eukaryotic microbes. The cell walls of eukaryotic microbes are typically composed of a single ingredient, like the cellulose found in algal cell walls or the chitin in fungal cell walls.

The bacterial cell wall performs several functions as well, in addition to providing overall strength to the cell. It also helps maintain the cell shape, which is important for how the cell will grow, reproduce, obtain nutrients, and move.

It protects the cell from osmotic lysis, as the cell moves from one environment to another or transports in nutrients from its surroundings. Since water can freely move across both the cell membrane and the cell wall, the cell is at risk for an osmotic imbalance, which could put pressure on the relatively weak plasma membrane. Studies have actually shown that the internal pressure of a cell is similar to the pressure found inside a fully inflated car tire. That is a lot of pressure for the plasma membrane to withstand! The cell wall can keep out certain molecules, such as toxins, particularly for gram negative bacteria. And lastly, the bacterial cell wall can contribute to the pathogenicity or disease –causing ability of the cell for certain bacterial pathogens.

STRUCTURE OF PEPTIDOGLYCAN

Let us start with peptidoglycan, since it is an ingredient that both bacterial cell walls have in common.

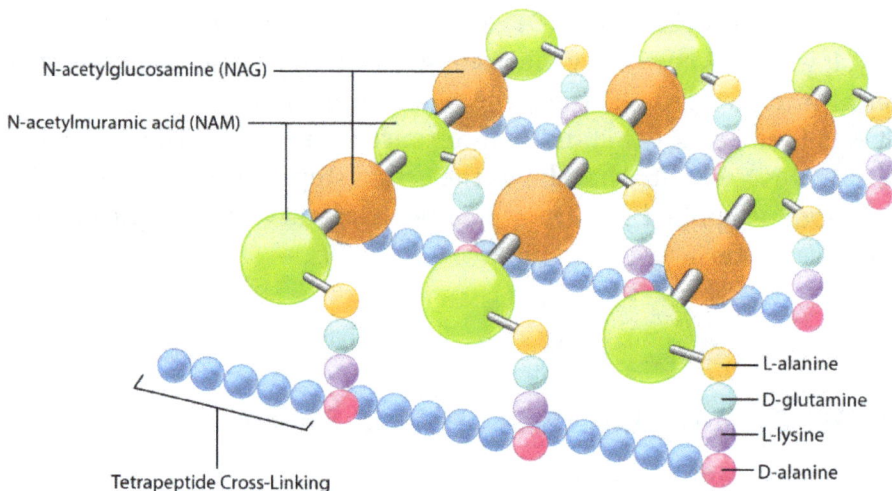

N-acetylglucosamine (NAG)

N-acetylmuramic acid (NAM)

L-alanine
D-glutamine
L-lysine
D-alanine

Tetrapeptide Cross-Linking

Peptidoglycan Structure.

Peptidoglycan is a polysaccharide made of two glucose derivatives, N-acetylglucosamine (NAG) and N-acetylmuramic acid (NAM), alternated in long chains. The chains are cross-linked to one another by a tetrapeptide that extends off the NAM sugar unit, allowing a lattice-like structure to form. The four amino acids that compose the tetrapeptide are: L-alanine, D-glutamine, L-lysine or meso-diaminopimelic acid (DPA), and D-alanine. Typically only the L-isomeric form of amino acids are utilized by cells but the use of the mirror image D-amino acids provides protection from proteases that might compromise the integrity of the cell wall by attacking the peptidoglycan. The tetrapeptides can be directly cross-linked to one another, with the D-alanine on one tetrapeptide binding to the L-lysine/ DPA on another tetrapeptide. In many gram positive bacteria there is a cross-bridge of five amino acids such as glycine (peptide interbridge) that serves to connect one tetrapeptide to another. In either case the cross-linking serves to increase the strength of the overall structure, with more strength derived from complete cross-linking, where every tetrapeptide is bound in some way to a tetrapeptide on another NAG-NAM chain.

While much is still unknown about peptidoglycan, research in the past ten years suggests that peptidoglycan is synthesized as a cylinder with a coiled substructure, where each coil is cross-linked to the coil next to it, creating an even stronger structure overall.

GRAM POSITIVE CELL WALLS

The cell walls of gram positive bacteria are composed predominantly of peptidoglycan. In fact, peptidoglycan can represent up to 90% of the cell wall, with layer after layer forming around the cell membrane. The NAM tetrapeptides are typically cross-linked with a peptide interbridge and complete cross-linking is common. All of this combines together to create an incredibly strong cell wall.

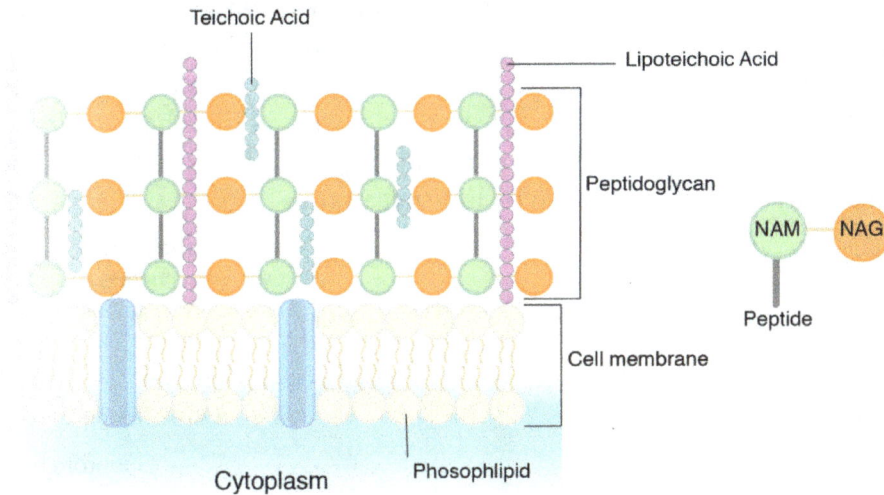

Gram Positive Bacteria Cell Wall

Since peptidoglycan is relatively porous, most substances can pass through the gram positive cell wall with little difficulty. But some nutrients are too large, requiring the cell to rely on the use of exoenzymes. These extracellular enzymes are made within the cell's cytoplasm and then secreted past the cell membrane, through the cell wall, where they function outside of the cell to break down large macromolecules into smaller components.

GRAM NEGATIVE CELL WALLS

The cell walls of gram negative bacteria are more complex than that of gram positive bacteria, with more ingredients overall. They do contain peptidoglycan as well, although only a couple of layers, representing 5-10% of the total cell wall. What is most notable about the gram negative cell wall is the presence of a plasma membrane located outside of the peptidoglycan layers, known as the outer membrane. This makes up the bulk of the gram negative cell wall. The outer membrane is composed of a lipid bilayer, very similar in composition to the cell membrane with polar heads, fatty acid tails, and integral proteins. It differs from the cell membrane by the presence of large molecules known as lipopolysaccharide (LPS), which are anchored into the outer membrane and project from the cell into the environment. LPS is made up of three different components: 1) the O-antigen or O-polysaccharide, which represents the outermost part of the structure , 2) the core polysaccharide, and 3) lipid A, which anchors the LPS into the outer membrane. LPS is known to serve many different functions for the cell, such as contributing to the net negative charge for the cell, helping to stabilize the outer membrane, and providing protection from certain chemical substances by physically blocking access to other parts of the cell wall. In addition, LPS plays a role in the host response to pathogenic gram negative bacteria. The O-antigen triggers an immune response in an infected host, causing the generation of antibodies specific to that part of LPS (think of E. coli O157). Lipid A acts as a toxin, specifically an endotoxin, causing general symptoms of illness such as fever and diarrhea. A large amount of lipid A released into the bloodstream can trigger endotoxic shock, a body-wide inflammatory response which can be life-threatening.

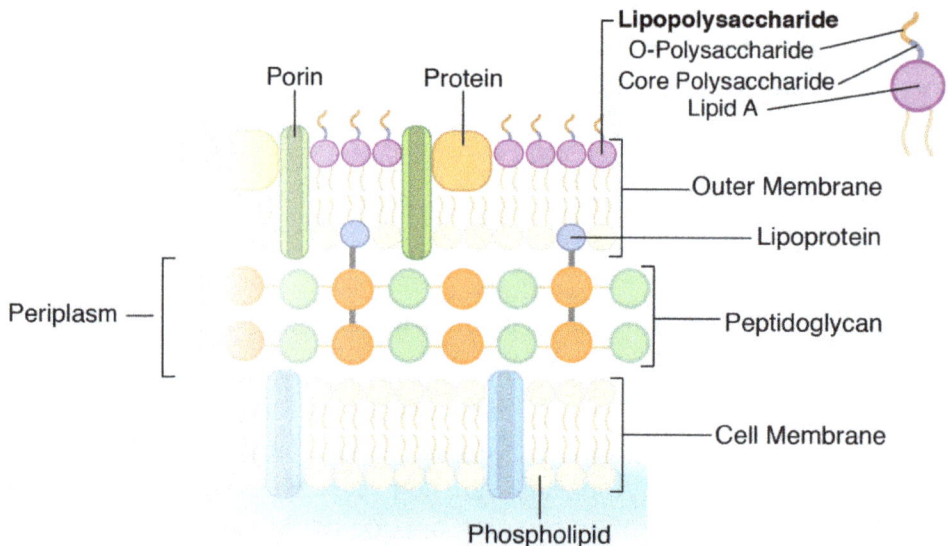

Gram Negetive Bacteria Cell Wall

The outer membrane does present an obstacle for the cell. While there are certain molecules it would like to keep out, such as antibiotics and toxic chemicals, there are nutrients that it would like to let in and the additional lipid bilayer presents a formidable barrier. Large molecules are broken down by enzymes, in order to allow them to get past the LPS. Instead of exoenzymes (like the gram positive bacteria), the gram negative bacteria utilize periplasmic enzymes that are stored in the periplasm. Where is the periplasm, you ask? It is the space located between the outer surface of the cell membrane and the inner surface of the outer membrane, and it contains the gram negative peptidoglycan. Once the periplasmic enzymes have broken nutrients down to smaller molecules that can get past the LPS, they still need to be transported across the outer membrane, specifically the lipid bilayer. Gram negative cells utilize porins, which are transmembrane proteins composed of a trimer of three subunits, which form a pore across the membrane. Some porins are non-specific and transport any molecule that fits, while some porins are specific and only transport substances that they recognize by use of a binding site. Once across the outer membrane and in the periplasm, molecules work their way through the porous peptidoglycan layers before being transported by integral proteins across the cell membrane.

The peptidoglycan layers are linked to the outer membrane by the use of a lipoprotein known as Braun's lipoprotein (good ol' Dr. Braun). At one end the lipoprotein is covalently bound to the peptidoglycan while the other end is embedded into the outer membrane via its polar head. This linkage between the two layers provides additional structural integrity and strength.

UNUSUAL AND WALL-LESS BACTERIA

Having emphasized the important of a cell wall and the ingredient peptidoglycan to both the gram positive and the gram negative bacteria, it does seem important to point out a few exceptions as well. Bacteria belonging to the phylum Chlamydiae appear to lack peptidoglycan, although their cell walls have a gram negative structure in all other regards (i.e. outer membrane, LPS, porin, etc). It has been suggested that they might be using a protein layer that functions in much the same way as peptidoglycan. This has an advantage to the cell in providing resistance to β-lactam antibiotics (such as penicillin), which attack peptidoglycan.

Bacteria belonging to the phylum Tenericutes lack a cell wall altogether, which makes them extremely susceptible to osmotic changes. They often strengthen their cell membrane somewhat by the addition of sterols, a substance usually associated with eukaryotic cell membranes. Many members of this phylum are pathogens, choosing to hide out within the protective environment of a host.

Foundations of Microbiology

Key Words

Cell wall, gram positive bacteria, gram negative bacteria, Gram stain, peptidoglycan, murein, osmotic lysis, N-acetylglucosamine (NAG), N-acetylmuramic acid (NAM), tetrapeptide, L-alanine, D-glutamine, L-lysine, meso-diaminopimelic acid (DPA), D-alanine, direct cross-link, peptide interbridge, complete cross-linking, teichoic acid, wall teichoic acid (WTA), lipoteichoic acid, exoenzymes, outer membrane, lipopolysaccharide (LPS), O-antigen or O-polysaccharide, core polysaccharide, lipid A, endotoxin, periplasmic enzymes, periplasm, porins, Braun's lipoprotein, Chlamydiae, Tenericutes, sterols.

05 | Internal Components of Bacteria

INTRODUCTION

We have already covered the main internal components found in all bacteria, namely, cytoplasm, the nucleoid, and ribosomes. Remember that bacteria are generally thought to lack organelles, those bilipid membrane-bound compartments so prevalent in eukaryotic cells (although some scientists argue that bacteria possess structures that could be thought of as simple organelles). But bacteria can be more complex, with a variety of additional internal components to be found that can contribute to their capabilities. Most of these components are cytoplasmic but some of them are periplasmic, located in the space between the cytoplasmic and outer membrane in gram negative bacteria.

CYTOSKELETON

It was originally thought that bacteria lacked a cytoskeleton, a significant component of eukaryotic cells. In the last 20 years, however, scientists have discovered bacterial filaments made of proteins that are analogues to the cytoskeletal proteins found in eukaryotes. It has also been determined that the bacterial cytoskeleton plays important roles in cell shape, cell division, and integrity of the cell wall.

FtsZ

FtsZ, homologous to the eukaryotic protein tubulin, forms a ring structure in the middle of the cell during cell division, attracting other proteins to the area in order to construct a septum that will eventually separate the two resulting daughter cells.

MreB

MreB, homologous to the eukaryotic protein actin, is found in bacillus and spiral-shaped bacteria and plays an essential role in cell shape formation. MreB assumes a helical configuration running the length of the cell and dictates the activities of the peptidoglycan-synthesis machinery, assuring a non-spherical shape.

Crescentin

Crescentin, homologous to the eukaryotic proteins lamin and keratin, is found in spiral-shaped bacteria with a single curve. The protein assembles lengthwise in the inner curvature of the cell, bending the cell into its final shape.

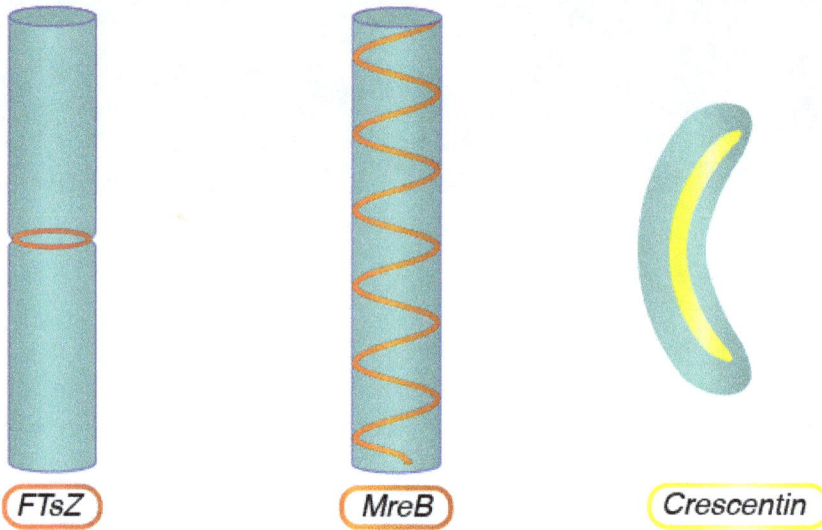

FTsZ   MreB   Crescentin

Cytoskeleton Structures.

INCLUSIONS

Bacterial inclusions are generally defined as a distinct structure located either within the cytoplasm or periplasm of the cell. They can range in complexity, from a simple compilation of chemicals such as crystals, to fairly complex structures that start to rival that of the eukaryotic organelles, complete with a membranous external layer. Their role is often to store components as metabolic reserves for the cell when a substance is found in excess, but they can also play a role in motility and metabolic functions as well.

CARBON STORAGE

Carbon is the most common substance to be stored by a cell, since all cells are carbon based. In addition, carbon compounds can often be broken down quickly by the cell, so they can serve as energy sources as well. One of the simplest and most common inclusions for carbon storage is glycogen, in which glucose units are linked together in a multi-branching polysaccharide structure.

Another common way for bacteria to store carbon is in the form of poly-β-hydroxybutyrate (PHB), a granule that forms when β-hydroxybutyric acid units aggregate together. This lipid is very plastic-like in composition, leading some scientists to investigate the possibility of using them as a biodegradeable plastic. The PHB granules actually have a shell composed of both protein and a small amount of phospholipid. Both glycogen and PHB are formed when there is an excess of carbon and then broken down by the cell later for both carbon and energy.

INORGANIC STORAGE

Often bacteria need something other than carbon, either for synthesis of cell components or as an alternate energy reserve. Polyphosphate granules allow for the accumulation of inorganic phosphate (PO_4^{-3}), where the phosphate can be used to make nucleic acid (remember the sugar-phosphate backbone?) or ATP (adenosine triphosphate, of course).

Other cells need sulfur as a source of electrons for their metabolism and will store excess sulfur in the form of sulfur globules, which result when the cell oxidizes hydrogen sulfide (H_2S) to elemental sulfur (S^0), resulting in the formation of refractile inclusions.

NON-STORAGE FUNCTIONS

There are times when a bacterium needs to do something beyond simple storage of organic or inorganic compounds for use in metabolism and there are inclusions to help with these non-storage functions. One such example is gas vacuoles, which are used by the cell to control buoyancy in a water column, providing the cell with some control over where it is in the environment. It is a limited form of motility, on the vertical axis only. Gas vacuoles are composed of conglomerations of gas vesicles, cylindrical structures that are both hollow and rigid. The gas vesicles are freely permeable to all types of gases by passive diffusion and can be quickly constructed or collapsed, as needed by the cell to ascend or descend.

Magnetosomes are inclusions that contain long chains of magnetite (Fe_3O_4), which are used by the cell as a compass in geomagnetic fields, for orientation within their environment. Magnetotactic bacteria are typically microaerophilic, preferring an environment with a lower level of oxygen than the atmosphere. The magenetosome allows the cells to locate the optimum depth for their growth. Magenetosomes have a true lipid bilayer, reminiscent of eukaryotic organelles, but it is actually an invagination of the cell's plasma membrane that has been modified with specific proteins.

MICROCOMPARTMENTS

Bacterial microcompartments (BMCs) are unique from other inclusions by virtue of their structure and functionality. They are icosahedral in shape and composed of a protein shell made up of various proteins in the BMC family. While their exact role varies, they all participate in functions beyond simple storage of substances. These compartments provide both a location and the substances (usually enzymes) necessary for particular metabolic activities.

The best studied example of a BMC is the carboxysome, which are found in many CO_2-fixing bacteria. Carboxysomes contain the enzyme ribulose-1,5-bisphosphate carboxylase (luckily it is also known as RubisCO), which plays a crucial role in converting CO2 into sugar. The carboxysome also plays a role in the concentration of CO2, thus ensuring that the components necessary for CO2-fixation are all in the same place at the same time.

ANAMMOXOSOME

The anammoxosome is a large membrane-bound compartment found in bacterial cells capable of carrying out the anammox reaction (anaerobic ammonium oxidation), where ammonium (NH4+) and nitrite (NO2-) are converted to dinitrogen gas (N2). The process is performed as a way for the cell to get energy, using ammonium as an electron donor and nitrite as an electron accept, with the resulting production of nitrogen gas. This chemical conversion of nitrogen is important for the nitrogen cycle.

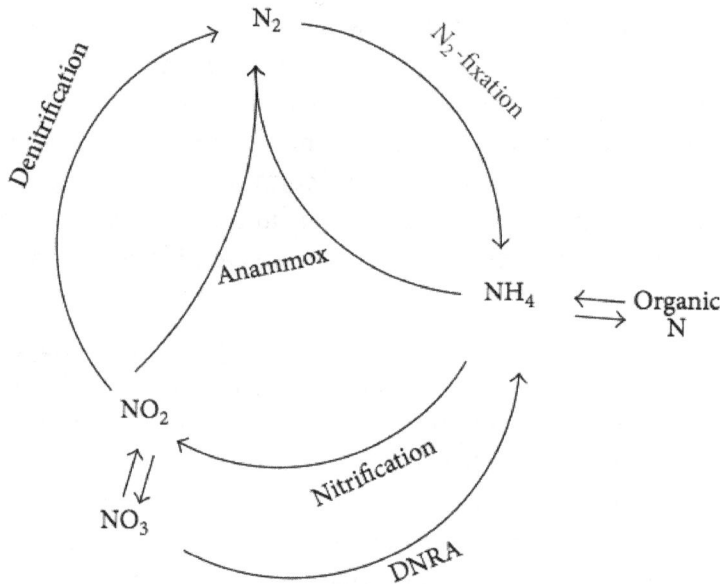

Nitrogen Cycle

CHLOROSOME

Found in some phototrophic bacteria, a chlorosome is a highly efficient structure for capturing low light intensities. Lining the inside perimeter of the cell membrane, each chlorosome can contain up to 250,000 bacteriochlorophyll molecules, arranged in dense arrays. Harvested light is transferred to reaction centers in the cell membrane, allowing the conversion from light energy to chemical energy in the form of ATP. The chlorosome is bounded by a lipid monolayer.

PLASMID

A plasmid is an extrachromosomal piece of DNA that some bacteria have, in addition to the genetic material found in the nucleoid. It is composed of double-stranded DNA and is typically circular, although linear plasmids have been found. Plasmids are described as being "non-essential" to the cell, where the cell can function normally in their absence. But while plasmids have only a few genes, they can confer important capabilities for the cell, such as antibiotic resistance. Plasmids replicate independently of the cell and can be lost (known as curing), either spontaneously or due to exposure to adverse conditions, such as UV light, thymine starvation, or growth above optimal conditions. Some plasmids, known as episomes, can be integrated into the cell chromosome where the genes will be replicated during cell division.

ENDOSPORE

Then there is the endospore, a marvel of bacterial engineering. This is located under the heading "bacterial internal components," but it is important to note that an endospore isn't an internal or external structure so much as a conversion of the cell into an alternate form. Cells start out as a vegetative cell, doing all the things a cell is supposed to do (metabolizing, reproducing, mowing the lawn...). If they get exposed to hostile conditions (dessication, high heat, an angry neighbor...) and they have the ability, they might convert from a vegetative cell into an endospore. The endospore is actually formed within the vegetative cell (doesn't that make it an internal structure?) and then the vegetative cell lyses, releasing the endospore (does that make it an external structure?).

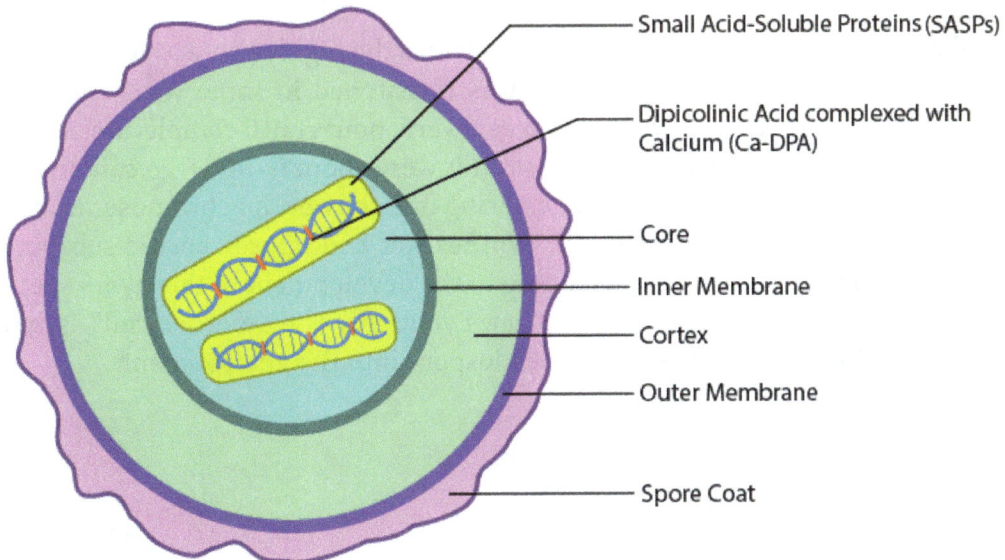

Small Acid-Soluble Proteins (SASPs)

Dipicolinic Acid complexed with Calcium (Ca-DPA)

Core

Inner Membrane

Cortex

Outer Membrane

Spore Coat

Endospore Layers

Endospores are only formed by a few gram positive genera and provide the cell with resistance to a wide variety of harsh conditions, such as starvation, extremes in temperature, exposure to drying, UV light, chemicals, enzymes, and radiation. While the vegetative cell is the active form for bacterial cells (growing, metabolizing, etc), the endospore can be thought of as a dormant form of the cell. It allows for survival of adverse conditions, but it does not allow the cell to grow or reproduce.

STRUCTURE

In order to be so incredibly resistant to so many different substances and environmental conditions, many different layers are necessary. The bacterial endospore has many different layers, starting with a core in the center. The core is the location of the nucleoid, ribosomes, and cytoplasm of the cell, in an extremely dehydrated form. It typically contains only 25% of the water found in a vegetative cell, increasing heat resistance. The DNA is further protected by the presence of small acid-soluble proteins (SASPs), which stabilize the DNA and protect it from degradation. DNA stabilization is increased by the presence of dipicolinic acid complexed with calcium (Ca-DPA), which inserts between the DNA bases. The core is wrapped in an inner membrane that provides a permeability barrier to chemicals, which is then surrounded by the cortex, a thick layer consisting of peptidoglycan with less cross-linking than is found in the vegetative cell. The cortex is wrapped in an outer membrane. Lastly are several spore coats made of protein, which provide protection from environmental stress such as chemicals and enzymes.

SPORULATION: CONVERSION FROM VEGETATIVE CELL TO ENDOSPORE

Sporulation, the conversion of vegetative cell into the highly protective endospore, typically occurs when the cell's survival is threatened in some way. The actual process is very complex and typically takes several hours until completion. Initially the sporulating cells replicates its DNA, as if it were about to undergo cell division. A septum forms asymmetrically, sequestering one copy of the chromosome at one end of the cell (called the forespore). Synthesis of endospore-specific substances occurs, altering the forespore and leading to the development of the layers specific for an endospore, as well as dehydration. Eventually the "mother cell" is lyses, allowing for the release of the mature endospore into the environment.

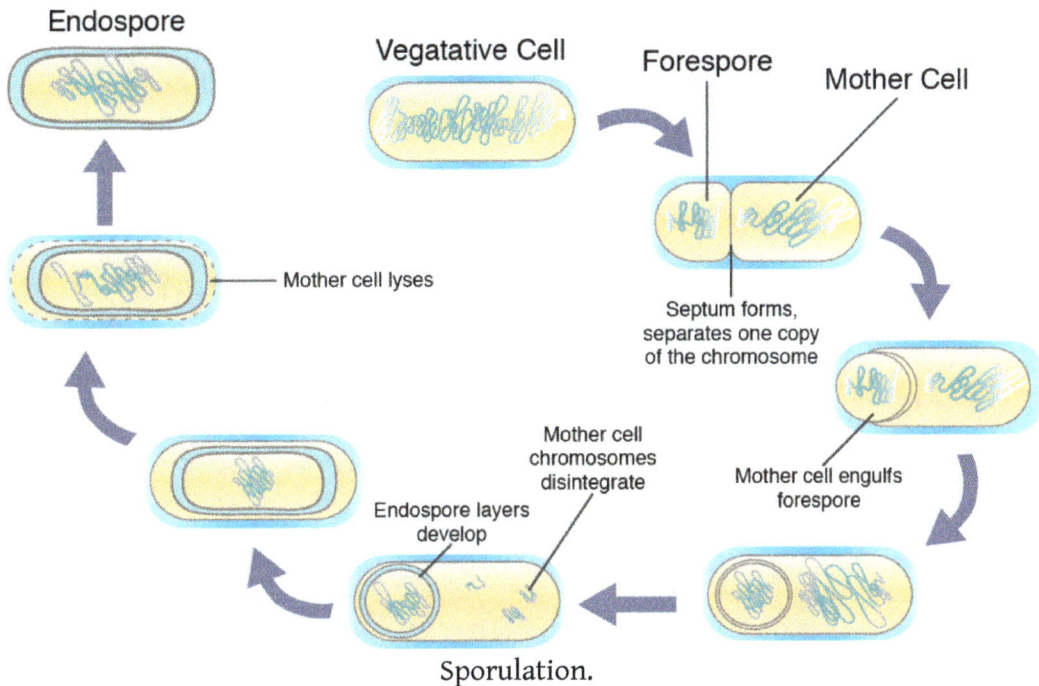

Sporulation.

CONVERSION FROM ENDOSPORE TO VEGETATIVE CELL

The endospore remains dormant until environmental conditions improve, causing a chemical change that initiates gene expression. There are three distinct stages in the conversion from an endospore to the metabolically active vegetative cells: 1) activation, a preparation step that can be initiated by the application of heat; 2) germination, when the endospore becomes metabolically active and begins to take on water; 3) outgrowth, when the vegetative cell fully emerges from the endospore shell.

Key Words
Cytoskeleton, FtsZ, tubulin, MreB, actin, crescentin, lamin, keratin, inclusion, glycogen, poly-β-hydroxybutyrate (PHB), polyphosphate granule, sulfur globule, gas vacuole, gas vesicle, magnetosome, microaerophilic, microcompartment, bacterial microcompartments (BMCs), carboxysome, ribulose-1,5-bisphosphate carboxylase, RubisCO, anammoxosome, anammox reaction, chlorosome, plasmid, curing, episome, endospore, vegetative cell, core, small acid-soluble proteins (SASPs), dipicolinic acid, Ca-DPA, inner membrane, cortex, outer membrane, spore coat, sporulation, forespore, activation, germination, outgrowth.

06 | Surface Structures of Bacteria

LAYERS OUTSIDE THE CELL WALL

What have we learned so far, in terms of cell layers? All cells have a cell membrane. Most bacteria have a cell wall. But there are a couple of additional layers that bacteria may, or may not, have. These would be found outside of both the cell membrane and the cell wall, if present.

CAPSULE

A bacterial capsule is a polysaccharide layer that completely envelopes the cell. It is well organized and tightly packed, which explains its resistance to staining under the microscope. The capsule offers protection from a variety of different threats to the cell, such as desiccation, hydrophobic toxic materials (i.e. detergents), and bacterial viruses. The capsule can enhance the ability of bacterial pathogens to cause disease and can provide protection from phagocytosis (engulfment by white blood cells known as phagocytes). Lastly, it can help in attachment to surfaces.

SLIME LAYER

A bacterial slime layer is similar to the capsule in that it is typically composed of polysaccharides and it completely surrounds the cell. It also offers protection from various threats, such as desiccation and antibiotics. It can also allow for adherence to surfaces. So, how does it differ from the capsule? A slime layer is a loose, unorganized layer that is easily stripped from the cell that made it, as opposed to a capsule which integrates firmly around the bacterial cell wall.

Some bacteria have a highly organized layer made of secreted proteins or glycoproteins that self-assemble into a matrix on the outer part of the cell wall. This regularly structured S-layer is anchored into the cell wall, although it is not considered to be officially part of the cell wall in bacteria. S-layers have very important roles for the bacteria that have them, particularly in the areas of growth and survival, and cell integrity.

Capsule

Slime Layer

S-LAYER

S layers help maintain overall rigidity of the cell wall and surface layers, as well as cell shape, which are important for reproduction. S layers protect the cell from ion/pH changes, osmotic stress, detrimental enzymes, bacterial viruses, and predator bacteria. They can provide cell adhesion to other cells or surfaces. For pathogenic bacteria they can provide protection from phagocytosis.

STRUCTURES OUTSIDE THE CELL WALL

Bacteria can also have structures outside of the cell wall, often bound to the cell wall and/or cell membrane. The building blocks for these structures are typically made within the cell and then secreted past the cell membrane and cell wall, to be assembled on the outside of the cell.

Fimbriae (sing. fimbria)

Fimbriae are thin filamentous appendages that extend from the cell, often in the tens or hundreds. They are composed of pilin proteins and are used by the cell to attach to surfaces. They can be particularly important for pathogenic bacteria, which use them to attach to host tissues.

Pili (sing. pilus)

Pili are very similar to fimbriae (some textbooks use the terms interchangeably) in that they are thin filamentous appendages that extend from the cell and are made of pilin proteins. Pili can be used for attachment as well, to both surfaces and host cells, such as the Neisseria gonorrhea cells that use their pili to grab onto sperm cells, for passage to the next human host. So, why would some researchers bother differentiating between fimbriae and pili?

Pili are typically longer than fimbriae, with only 1-2 present on each cell, but that hardly seems enough to set the two structures apart. It really boils down to the fact that a few specific pili participate in functions beyond attachment. The conjugative pili participate in the process known as conjugation, which allows for the transfer of a small piece of DNA from a donor cell to a recipient cell. The type IV pili play a role in an unusual type of motility known as twitching motility, where a pilus attaches to a solid surface and then contracts, pulling the bacterium forward in a jerky motion.

Flagella (sing. flagellum)

Bacterial motility is typically provided by structures known as flagella. The bacterial flagellum differs in composition, structure, and mechanics from the eukaryotic flagellum, which operates as a flexible whip-like tail utilizing microtubules that are powered by ATP. The bacterial flagellum is rigid in nature, operates more like the propeller on a boat, and is powered by energy from the proton motive force.

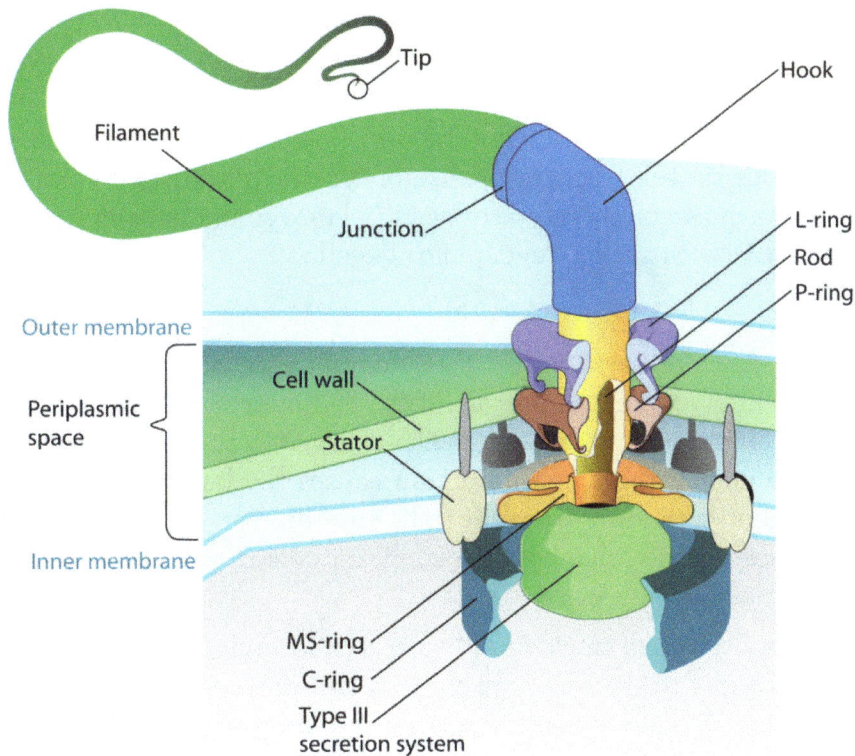

Flagellum base diagram

There are three main components to the bacterial flagellum:

- ◉ The filament – a long thin appendage that extends from the cell surface. The filament is composed of the protein flagellin and is hollow. Flagellin proteins are transcribed in the cell cytoplasm and then transported across the cell membrane and cell wall. A bacterial flagellar filament grows from its tip (unlike the hair on your head), adding more and more flagellin units to extend the length until the correct size is reached. The flagellin units are guided into place by a protein cap.

- ◉ The hook – this is a curved coupler that attaches the filament to the flagellar motor.

- ◉ The motor – a rotary motor that spans both the cell membrane and the cell wall, with additional components for the gram negative outer membrane. The motor has two components: the basal body, which provides the rotation, and the stator, which provides the torque necessary for rotation to occur. The basal body consists of a central shaft surrounded by protein rings, two in the gram positive bacteria and four in the gram negative bacteria. The stator consists of Mot proteins that surround the ring(s) embedded within the cell membrane.

BACTERIAL MOVEMENT

Bacterial movement typically involves the use of flagella, although there are a few other possibilities as well (such as the use of type IV pili for twitching motility). But certainly the most common type of bacterial movement is swimming, which is accomplished with the use of a flagellum or flagella.

Swimming

Rotation of the flagellar basal body occurs due to the proton motive force, where protons that accumulate on the outside of the cell membrane are driven through pores in the Mot proteins, interacting with charges in the ring proteins as they pass across the membrane. The interaction causes the basal body to rotate and turns the filament extending from the cell. Rotation can occur at 200-1000 rpm and result in speeds of 60 cell lengths/second (for comparison, a cheetah moves at a maximum rate of 25 body lengths/second).

Rotation can occur in a clockwise (CW) or a counterclockwise (CCW) direction, with different results to the cell. A bacterium will move forward, called a "run," when there is a CCW rotation, and reorient randomly, called a "tumble," when there is a CW rotation.

Corkscrew Motility

Some spiral-shaped bacteria, known as the Spirochetes, utilize a corkscrew-motility due to their unusual morphology and flagellar conformation. These gram negative bacteria have specialized flagella that attach to one end of the cell, extend back through the periplasm and then attach to the other end of the cell. When these endoflagella rotate they put torsion on the entire cell, resulting in a flexing motion that is particularly effective for burrowing through viscous liquids.

Gliding Motility

Gliding motility is just like it sounds, a slower and more graceful movement than the other forms covered so far. Gliding motility is exhibited by certain filamentous or bacillus bacteria and does not require the use of flagella. It does require that the cells be in contact with a solid surface, although more than one mechanism has been identified. Some cells rely on slime propulsion, where secreted slime propels the cell forward, where other cells rely on surface layer proteins to pull the cell forward.

Chemotaxis

Now that we have covered the basics of the bacterial flagellar motor and mechanics of bacterial swimming, let us combine the two topics to talk about chemotaxis or any other type of taxes (just not my taxes). Chemotaxis refers to the movement of an organism towards or away from a chemical. You can also have phototaxis, where an organism is responding to light. In chemotaxis, a favorable substance (such as a nutrient) is referred to as an attractant, while a substance with an adverse effect on the cell (such as a toxin) is referred to as a repellant. In the absence of either an attractant or a repellant a cell will engage in a "random walk," where it alternates between tumbles and runs, in the end getting nowhere in particular. In the presence of a gradient of some type, the movements of the cell will become biased, resulting over time in the movement of the bacterium towards an attractant and away from any repellants. How does this happen?

First, let us cover how a bacterium knows which direction to go. Bacteria rely on protein receptors embedded within their membrane, called chemoreceptors, which bind specific molecules. Binding typically results in methylation or phosphorylation of the chemoreceptor, which triggers an elaborate protein pathway that eventually impacts the rotation of the flagellar motor. The bacteria engage in temporal sensing, where they compare the concentration of a substance with the concentration obtained just a few seconds (or microseconds) earlier. In this way they gather information about the orientation of the concentration gradient of the substance. As a bacterium moves closer to the higher concentrations of an attractant, runs (dictated by CCW flagellar rotation) become longer, while tumbling (dictated by CW flagellar rotation) decreases. There will still be times that the bacterium will head off in the wrong direction away from an attractant since tumbling results in a random reorientation of the cell, but it won't head in the wrong direction for very long. The resulting "biased random walk" allows the cell to quickly move up the gradient of an attractant (or move down the gradient of a repellant).

Bacterial movement

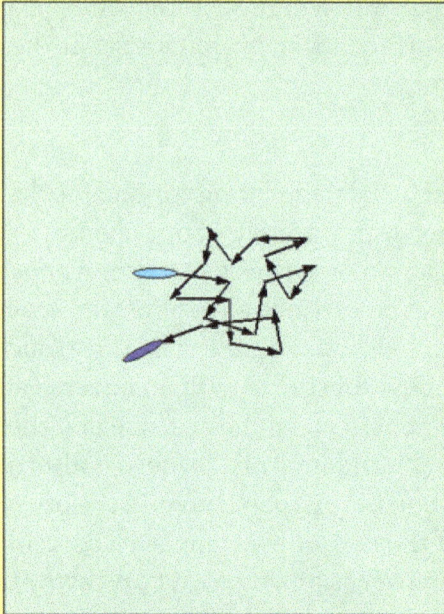

Random walk
(No stimulus)

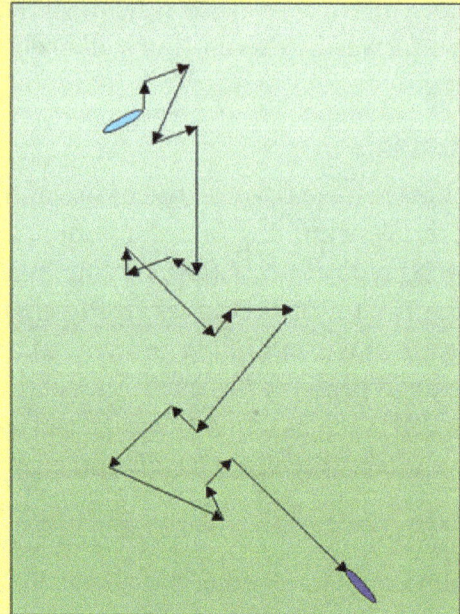

Chemotaxis
(Positive Stimulus)

Brudersohn
Wikimedia Commons

Bacterial Movement

KEY WORDS

Capsule, slime layer, S-layer, fimbriae/fimbria, pilin, pili/pilus, conjugative pili, conjugation, type IV pili, twitching motility, flagella/flagellum, filament, flagellin, hook, motor, basal body, stator, Mot proteins, swimming, clockwise (CW), counterclockwise (CCW), run, tumble, Spirochetes, corkscrew-motility, endoflagella, gliding motility, chemotaxis, phototaxis, attractant, repellant, random walk, chemoreceptors, temporal sensing, biased random walk.

The Archaea are a group of organisms that were originally thought to be bacteria (which explains the initial name of "archaeabacteria"), due to their physical similarities. More reliable genetic analysis revealed that the Archaea are distinct from both Bacteria and Eukaryotes, earning them their own domain in the Three Domain Classification originally proposed by Woese in 1977, alongside the Eukarya and the Bacteria.

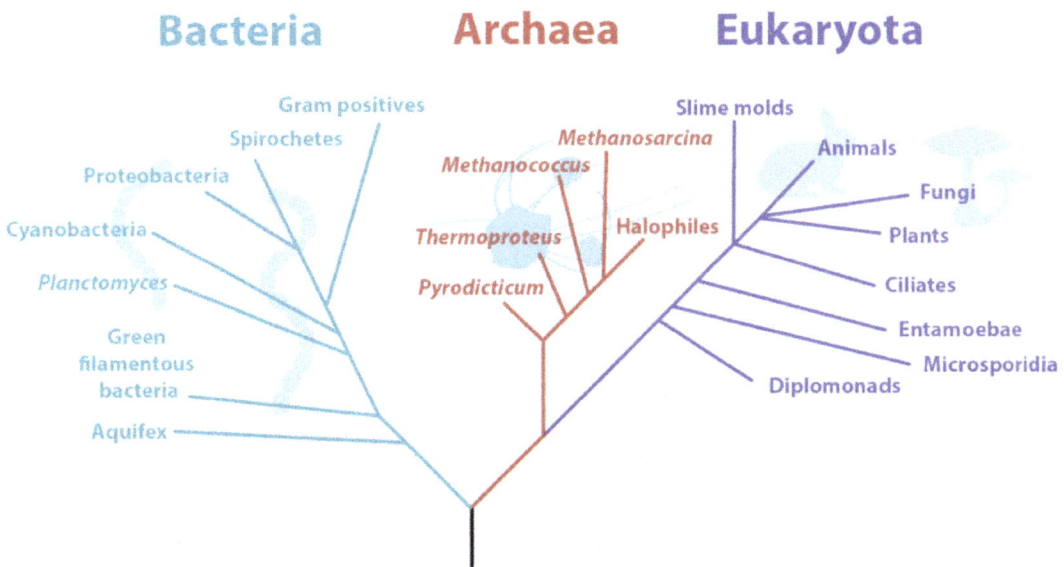

Bacteria   **Archaea**   **Eukaryota**

Gram positives
Spirochetes
Proteobacteria
Cyanobacteria
Planctomyces
Green filamentous bacteria
Aquifex

Methanosarcina
Methanococcus
Thermoproteus
Pyrodicticum
Halophiles

Slime molds
Animals
Fungi
Plants
Ciliates
Entamoebae
Microsporidia
Diplomonads

Phylogenetic Tree of Life.

SIMILARITIES TO BACTERIA

So, why were the archaea originally thought to be bacteria? Perhaps most importantly, they lack a nucleus or other membrane-bound organelles, putting them into the prokaryotic category (if you are using the traditional classification scheme). Most of them are unicellular, they have 70S sized ribosomes, they are typically a few micrometers in size, and they reproduce asexually only. They are known to have many of the same structures that bacteria can have, such as plasmids, inclusions, flagella, and pili. Capsules and slime layers have been found but appear to be rare in archaea.

While archaea were originally isolated from extreme environments, such as places high in acid, salt, or heat, earning them the name "extremophiles," they have more recently been isolated from all the places rich with bacteria: surface water, the ocean, human skin, soil, etc.

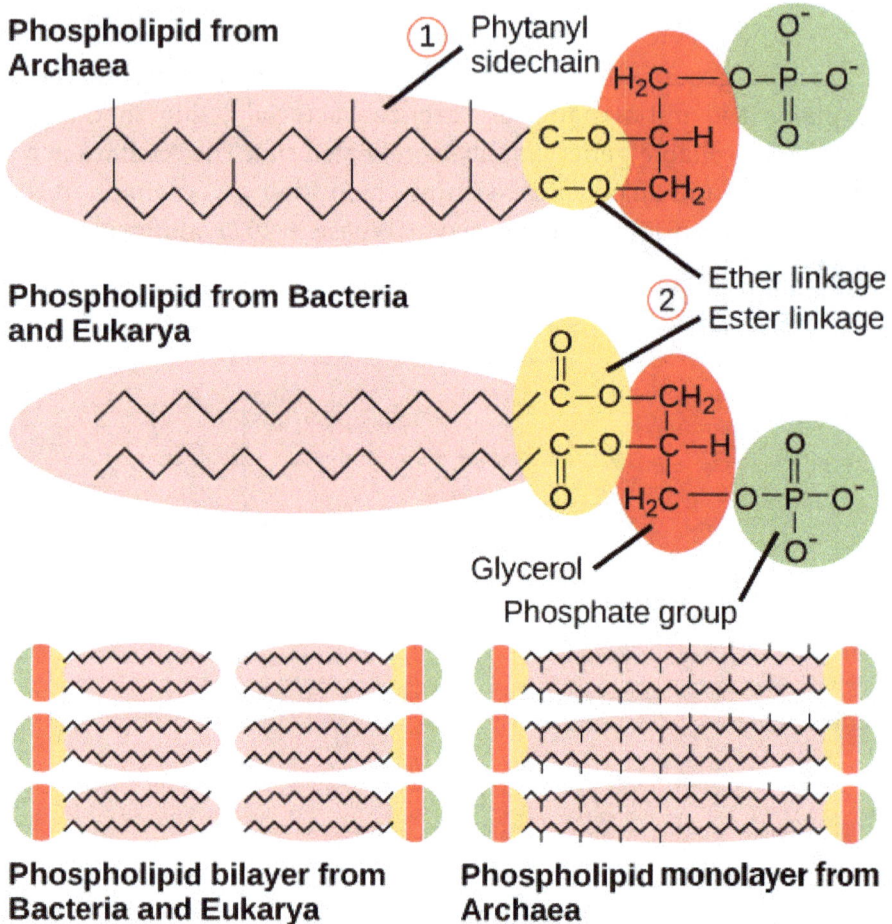

Comparison of Plasma Membrane Lipid Between Bacteria and Archaea. OpenStax, Structure of Prokaryotes. OpenStax CNX

KEY DIFFERENCES

Plasma Membrane

There are several characteristics of the plasma membrane that are unique to Archaea, setting them apart from other domains. One such characteristic is chirality of the glycerol linkage between the phopholipid head and the side chain. In archaea it is in the L-isomeric form, while bacteria and eukaryotes have the D-isomeric form. A second difference is the presence of an ether-linkage between the glycerol and the side chain, as opposed to the ester-linked lipids found in bacteria and eukaryotes. The ether-linkage provides more chemical stability to the membrane. A third and fourth difference are associated with the side chains themselves, unbranched fatty acids in bacteria and eukaryotes, while isoprenoid chains are found in archaea. These isoprenoid chains can have branching side chains.

Lastly, the plasma membrane of Archaea can be found as monolayers, where the isoprene chains of one phospholipid connect with the isoprene chains of a phospholipid on the opposite side of the membrane. Bacteria and eukaryotes only have lipid bilayers, where the two sides of the membrane remain separated.

CELL WALL

Like bacteria, the archaeal cell wall is a semi-rigid structure designed to provide protection to the cell from the environment and from the internal cellular pressure. While the cell walls of bacteria typically contain peptidoglycan, that particular chemical is lacking in archaea. Instead, archaea display a wide variety of cell wall types, adapted for the environment of the organism. Some archaea lack a cell wall altogether.

While it is not universal, a large number of Archaea have a proteinaceous S-layer that is considered to be part of the cell wall itself (unlike in Bacteria, where an S-layer is a structure in addition to the cell wall). For some Archaea the S-layer is the only cell wall component, while in others it is joined by additional ingredients (see below). The archaeal S-layer can be made of either protein or glycoprotein, often anchored into the plasma membrane of the cell. The proteins form a two-dimensional crystalline array with a smooth outer surface. A few S-layers are composed of two different S-layer proteins.

While archaea lack peptidoglycan, a few contain a substance with a similar chemical structure, known as pseudomurein. Instead of NAM, it contains N-acetylalosaminuronic acid (NAT) linked to NAG, with peptide interbridges to increase strength.

Methanochondroitin is a cell wall polymer found in some archaeal cells, similar in composition to the connective tissue component chondroitin, found in vertebrates.

Some archaea have a protein sheath composed of a lattice structure similar to an S-layer. These cells are often found in filamentous chains, however, and the protein sheath encloses the entire chain, as opposed to individual cells.

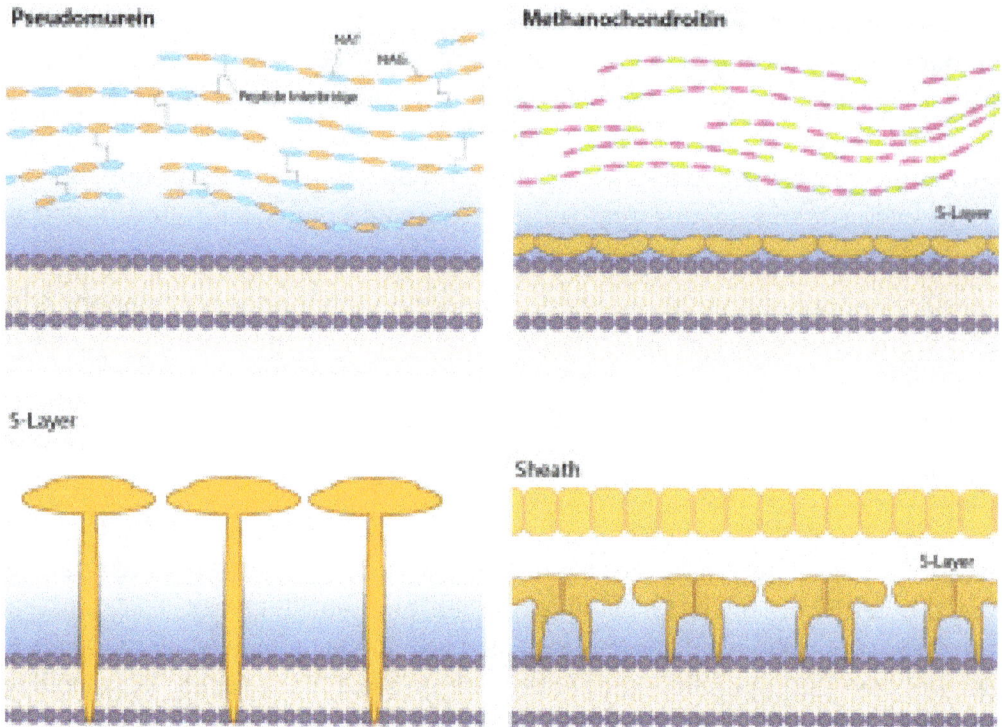

Cell Wall Structural Diversity.

RIBOSOMES

While archaea have ribosomes that are 70S in size, the same as bacteria, it was the rRNA nucleotide differences that provided scientists with the conclusive evidence to argue that archaea deserved a domain separate from the bacteria. In addition, archaeal ribosomes have a different shape than bacterial ribosomes, with proteins that are unique to archaea. This provides them with resistance to antibiotics that inhibit ribosomal function in bacteria.

STRUCTURES

Many of the structures found in bacteria have been discovered in archaea as well, although sometimes it is obvious that each structure was evolved independently, based on differences in substance and construction.

Cannulae

Cannulae, a structure unique to archaea, have been discovered in some marine archaeal strains. These hollow tube-like structures appear to connect cells after division, eventually leading to a dense network composed of numerous cells and tubes. This could serve as a means of anchoring a community of cells to a surface.

Hami (sing. hamus)

Another structure unique to archaea is the hamus, a long helical tube with three hooks at the far end. Hami appear to allow cells to attach both to one another and to surfaces, encouraging the formation of a community.

Pili (sing. pilus)

Pili have been observed in archaea, composed of proteins most likely modified from the bacterial pilin. The resulting tube-like structures have been shown to be used for attachment to surfaces.

Flagella (sing. flagellum)

The archaeal flagellum, while used for motility, differs so markedly from the bacterial flagellum that it has been proposed to call it an "archaellum," to differentiate it from its bacterial counterpart.

What is similar between the bacterial flagellum and the archaeal flagellum? Both are used for movement, where the cell is propelled by rotation of a rigid filament extending from the cell. After that the similarities end.

What are the differences? The rotation of an archaeal flagellum is powered by ATP, as opposed to the proton motive force used in bacteria. The proteins making up the archaeal flagellum are similar to the proteins found in bacterial pili, rather than the bacterial flagellum. The archaeal flagellum filament is not hollow so growth occurs when flagellin proteins are inserted into the base of the filament, rather than being added to the end. The filament is made up of several different types of flagellin, while just one type is used for the bacterial flagellum filament. Clockwise rotation pushes an archaeal cells forward, while counterclockwise rotation pulls an archaeal cell backwards. An alternation of runs and tumbles is not observed.

CLASSIFICATION

Currently there are two recognized phyla of archaea: Euryarchaeota and Proteoarchaeota. Several additional phyla have been proposed (Nanoarchaeota, Korarchaeota, Aigarchaeota, Lokiarchaeota), but have yet to be officially recognized, largely due to the fact that the evidence comes from environmental sequences only.

KEY WORDS
Archaea, L-isomeric form, D-isomeric form, ether-linkages, ester-linkages, isoprenoid chains, branching side chains, lipid monolayer, lipid bilayer, S-layer, pseudomurein, N-acetylalosaminuronic acid (NAT), methanochondroitin, protein sheath, cannulae, hamus/hami, pilus/pili, flagellum/flagella, archaellum, Euryarchaeota, Proteoarchaeota.

Viruses are typically described as obligate intracellular parasites, acellular infectious agents that require the presence of a host cell in order to multiply. Viruses that have been found to infect all types of cells – humans, animals, plants, bacteria, yeast, archaea, protozoa...some scientists even claim they have found a virus that infects other viruses! But that is not going to happen without some cellular help.

VIRUS CHARACTERISTICS

Viruses can be extremely simple in design, consisting of nucleic acid surrounded by a protein coat known as a capsid. The capsid is composed of smaller protein components referred to as capsomers. The capsid+genome combination is called a nucleocapsid.

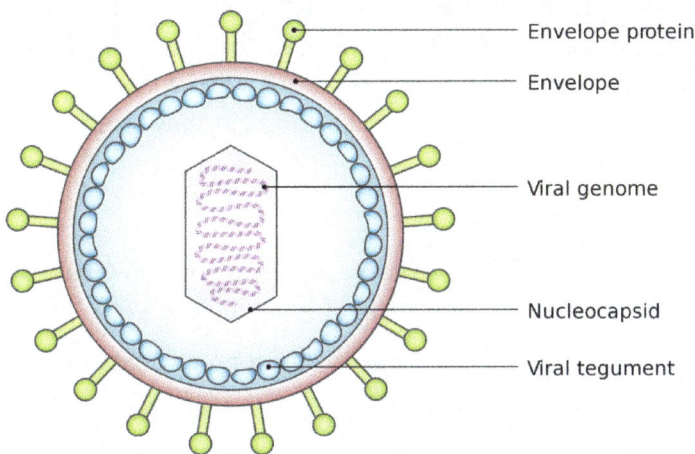

Envelope protein

Envelope

Viral genome

Nucleocapsid

Viral tegument

Virus Characteristics

Viruses can also possess additional components, with the most common being an additional membranous layer that surrounds the nucleocapsid, called an envelope. The envelope is actually acquired from the nuclear or plasma membrane of the infected host cell, and then modified with viral proteins called peplomers. Some viruses contain viral enzymes that are necessary for infection of a host cell and coded for within the viral genome. A complete virus, with all the components needed for host cell infection, is referred to as a virion.

VIRUS GENOME

While cells contain double-stranded DNA for their genome, viruses are not limited to this form. While there are dsDNA viruses, there are also viruses with single-stranded DNA (ssDNA), double-stranded RNA (dsRNA), and single-stranded RNA (ssRNA). In this last category, the ssRNA can either positive-sense (+ssRNA, meaning it can transcribe a message, like mRNA) or it can be negative-sense (-ssRNA, indicating that it is complementary to mRNA). Some viruses even start with one form of nucleic acid in the nucleocapsid and then convert it to a different form during replication.

VIRUS STRUCTURE

Viral nucleocapsids come in two basic shapes, although the overall appearance of a virus can be altered by the presence of an envelope, if present. Helical viruses have an elongated tube-like structure, with the capsomers arranged helically around the coiled genome. Icosahedral viruses have a spherical shape, with icosahedral symmetry consisting of 20 triangular faces. The simplest icosahedral capsid has 3 capsomers per triangular face, resulting in 60 capsomers for the entire virus. Some viruses do not neatly fit into either of the two previous categories because they are so unusual in design or components, so there is a third category known as complex viruses. Examples include the poxvirus with a brick-shaped exterior and a complicated internal structure, as well as bacteriophage with tail fibers attached to an icosahedral head.

VIRUS REPLICATION CYCLE

While the replication cycle of viruses can vary from virus to virus, there is a general pattern that can be described, consisting of five steps:

- ◉ Attachment – the virion attaches to the correct host cell.
- ◉ Penetration or Viral Entry – the virus or viral nucleic acid gains entrance into the cell.
- ◉ Synthesis – the viral proteins and nucleic acid copies are manufactured by the cells' machinery.
- ◉ Assembly – viruses are produced from the viral components.
- ◉ Release – newly formed virions are released from the cell.

ATTACHMENT

Outside of their host cell, viruses are inert or metabolically inactive. Therefore, the encounter of a virion to an appropriate host cell is a random event. The attachment itself is highly specific, between molecules on the outside of the virus and receptors on the host cell surface. This accounts for the specificity of viruses to only infect particular cell types or particular hosts.

PENETRATION OR VIRAL ENTRY

Many unenveloped (or naked) viruses inject their nucleic acid into the host cell, leaving an empty capsid on the outside. This process is termed penetration and is common with bacteriophage, the viruses that infect bacteria. With the eukaryotic viruses, it is more likely for the entire capsid to gain entrance into the cell, with the capsid being removed in the cytoplasm. An unenveloped eukaryotic virus often gains entry via endocytosis, where the host cell is compelled to engulf the capsid resulting in an endocytic vesicle, allowing the virus access to the cell contents. An enveloped eukaryotic virus gains entrance for its nucleocapsid through membrane fusion, where the viral envelope fuses with the host cell membrane, pushing the nucleocapsid past the cell membrane. If the entire nucleocapsid is brought into the cell then there is an uncoating process to strip away the capsid and release the viral genome.

SYNTHESIS

The synthesis stage is largely dictated by the type of viral genome, since genomes that differ from the cell's dsDNA genome can involve intricate viral strategies for genome replication and protein synthesis. Viral specific enzymes, such as RNA-dependent RNA polymerases, might be necessary for the replication process to proceed. Protein production is tightly controlled, to insure that components are made at the right time in viral development.

ASSEMBLY

The complexity of viral assembly depends upon the virus being made. The simplest virus has a capsid composed of 3 different types of proteins, which self-assembles with little difficulty. The most complex virus is composed of over 60 different proteins, which must all come together in a specific order. These viruses often employ multiple assembly lines to create the different viral structures and then utilize scaffolding proteins to put all the viral components together in an organized fashion.

RELEASE

The majority of viruses lyse their host cell at the end of replication, allowing all the newly formed virions to be released to the environment. Another possibility, common for enveloped viruses, is budding, where one virus is released from the cell at a time. The cell membrane is modified by the insertion of viral proteins, with the nucleocapsid pushing out through this modified portion of the membrane, allowing it to acquire an envelope.

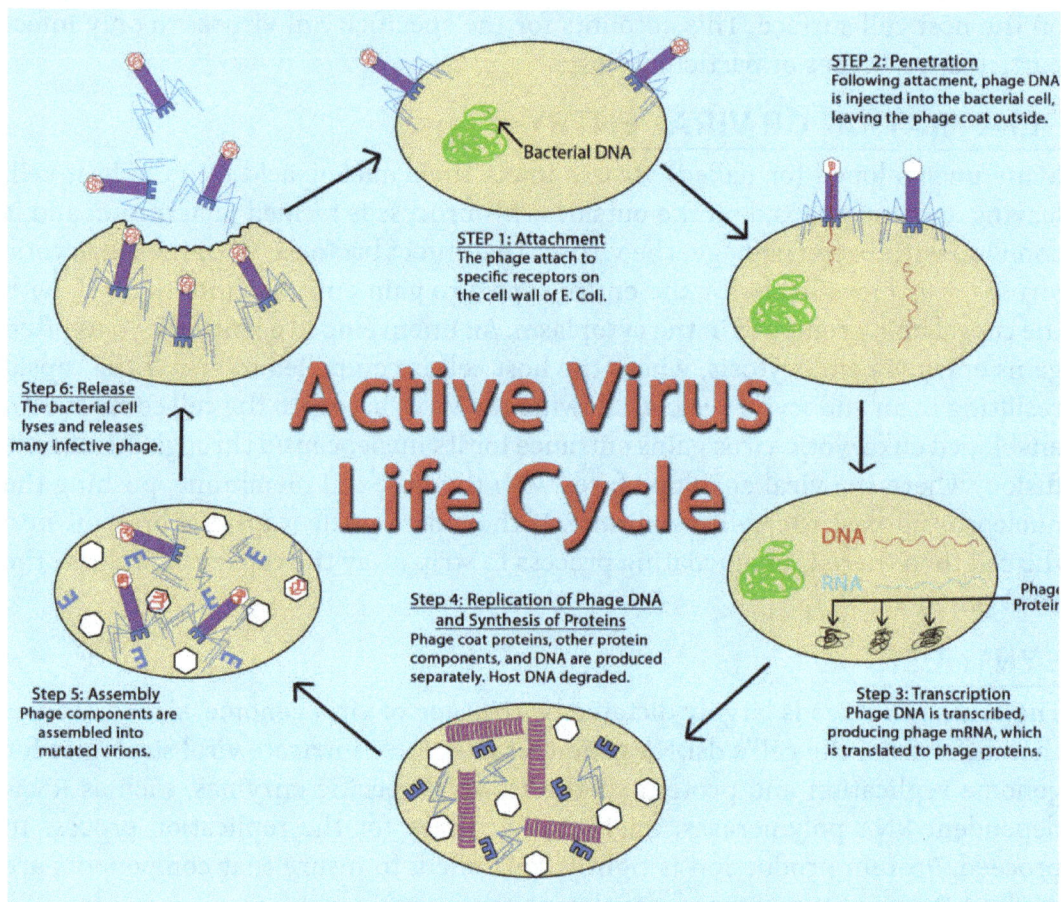

STEP 2: Penetration
Following attacment, phage DNA is injected into the bacterial cell, leaving the phage coat outside.

Bacterial DNA

STEP 1: Attachment
The phage attach to specific receptors on the cell wall of E. Coli.

Step 6: Release
The bacterial cell lyses and releases many infective phage.

Active Virus Life Cycle

Step 4: Replication of Phage DNA and Synthesis of Proteins
Phage coat proteins, other protein components, and DNA are produced separately. Host DNA degraded.

DNA

RNA

Phage Protein

Step 5: Assembly
Phage components are assembled into mutated virions.

Step 3: Transcription
Phage DNA is transcribed, producing phage mRNA, which is translated to phage proteins.

Active Virus Life Cycle

BACTERIOPHAGE

Viruses that infect bacteria are known as bacteriophage or phage. A virulent phage is one that always lyses the host cell at the end of replication, after following the five steps of replication described above. This is called the lytic cycle of replication.

There are also temperate phage, viruses that have two options regarding their replication. Option 1 is to mimic a virulent phage, following the five steps of replication and lysing the host cell at the end, referred to as the lytic cycle. But temperate phage differ from virulent phage in that they have another choice: Option 2, where they remain within the host cell without destroying it. This process is known as lysogeny or the lysogenic cycle of replication.

A phage employing lysogeny still undergoes the first two steps of a typical replication cycle, attachment and penetration. Once the viral DNA has been inserted into the cell it integrates with the host DNA, forming a prophage. The infected bacterium is referred to as a lysogen or lysogenic bacterium. In this state, the virus enjoys a stable relationship with its host, where it does not interfere with host cell metabolism or reproduction. The host cell enjoys immunity from reinfection from the same virus.

Exposure of the host cell to stressful conditions (i.e. UV irradiation) causes induction, where the viral DNA excises from the host cell DNA. This event triggers the remaining steps of the lytic cycle, synthesis, maturation, and release, leading to lysis of the host cell and release of newly formed virions.

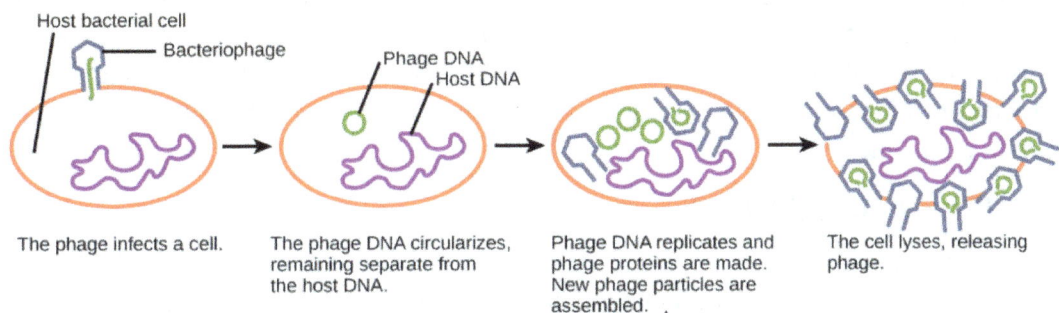

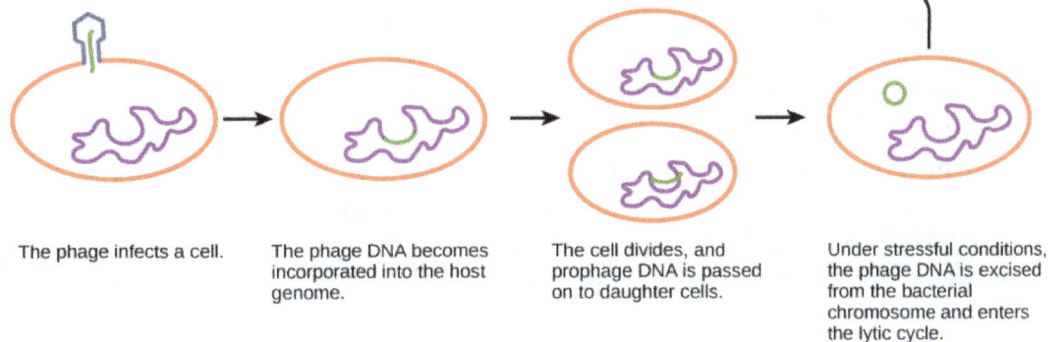

Lytic Cycle Versus Lysogenic Cycle of Replication. OpenStax, Virus Infections and Hosts. OpenStax CNX. Apr 11, 2013.

So, what dictates the replication type that will be used by a temperate phage? If there are plenty of host cells around, it is likely that a temperate phage will engage in the lytic cycle of replication, leading to a large increase in viral production. If host cells are scarce, a temperate phage is more likely to enter lysogeny, allowing for viral survival until host cell numbers increase. The same is true if the number of phage in an environment greatly outnumber the host cells, since lysogeny would allow for host cells numbers to rebound, ensuring long term viral survival.

Lysogens can experience a benefit from lysogeny as well, since it can result in lysogenic conversion, a situation where the development of a prophage leads to a change in the host's phenotype. One of the best examples of this is for the bacterium Corynebacterium diphtheriae, the causative agent of diphtheria. The diphtheria toxin that causes the disease is encoded within the phage genome, so only C. diphtheriae lysogens cause diphtheria.

EUKARYOTIC VIRUSES

Eukaryotic viruses can cause one of four different outcomes for their host cell. The most common outcome is host cell lysis, resulting from a virulent infection (essentially the lytic cycle of replication seen in phage). Some viruses can cause a latent infection, co-existing peacefully with their host cells for years (much like a temperate phage during lysogeny). Some enveloped eukaryotic viruses can also be released one at a time from an infected host cell, in a type of budding process, causing a persistent infection. Lastly, certain eukaryotic viruses can cause the host cell to transform into a malignant or cancerous cell, a mechanism known as transformation.

VIRUSES AND CANCER

There are many different causes of cancer, or unregulated cell growth and reproduction. Some known causes include exposure to certain chemicals or UV light. There are also certain viruses that have a known associated with the development of cancer. Such viruses are referred to as oncoviruses. Oncoviruses can cause cancer by producing proteins that bind to host proteins known as tumor suppressor proteins, which function to regulate cell growth and to initiate programmed cell death, if needed. If the tumor suppressor proteins are inactivated by viral proteins then cells grow out of control, leading to the development of tumors and metastasis, where the cells spread throughout the body.

Key Words
Virus, obligate intracellular parasite, capsid, bacteriophage, capsomere, nucleocapsid, envelope, peplomer, virion, dsDNA, ssDNA, dsRNA, +ssRNA, -ssRNA, helical viruses, icosahedral viruses, complex viruses, attachment, penetration, viral entry, synthesis, assembly, release, naked virus, endocytosis, membrane fusion, budding, bacteriophage, phage, virulent phage, lytic cycle, temperate phage, lysogeny, lysogenic cycle, prophage, lysogen, lysogenic bacterium, induction, lysogenic conversion, virulent infection, latent infection, persistent infection, transformation, oncovirus, tumor suppressor proteins.

09 | Microbial Development

INTRODUCTION

Provided with the right conditions (food, correct temperature, etc) microbes can grow very quickly. Depending on the situation, this could be a good thing for humans (yeast growing in wort to make beer) or a bad thing (bacteria growing in your throat causing strep throat). It's important to have knowledge of their growth, so we can predict or control their growth under particular conditions.

While growth for muticelluar organisms is typically measured in terms of the increase in size of a single organism, microbial growth is measured by the increase in population, either by measuring the increase in cell number or the increase in overall mass.

BACTERIAL DIVISION

Bacteria and archaea reproduce asexually only, while eukartyotic microbes can engage in either sexual or asexual reproduction. Bacteria and archaea most commonly engage in a process known as binary fission, where a single cell splits into two equally sized cells. Other, less common processes can include multiple fission, budding, and the production of spores.

The process begins with cell elongation, which requires careful enlargement of the cell membrane and the cell wall, in addition to an increase in cell volume. The cell starts to replicate its DNA, in preparation for having two copies of its chromosome, one for each newly formed cell. The protein FtsZ is essential for the formation of a septum, which initially manifests as a ring in the middle of the elongated cell. After the nucleoids are segregated to each end of the elongated cell, septum formation is completed, dividing the elongated cell into two equally sized daughter cells. The entire process or cell cycle can take as little as 20 minutes for an active culture of E. coli bacteria.

GROWTH CURVE

Since bacteria are easy to grow in the lab, their growth has been studied extensively. It has been determined that in a closed system or batch culture (no food added, no wastes removed) bacteria will grow in a predictable pattern, resulting in a growth curve composed of four distinct phases of growth: the lag phase, the exponential or log phase, the stationary phase, and the death or decline phase. Additionally, this growth curve can yield generation time for a particular organism – the amount of time it takes for the population to double.

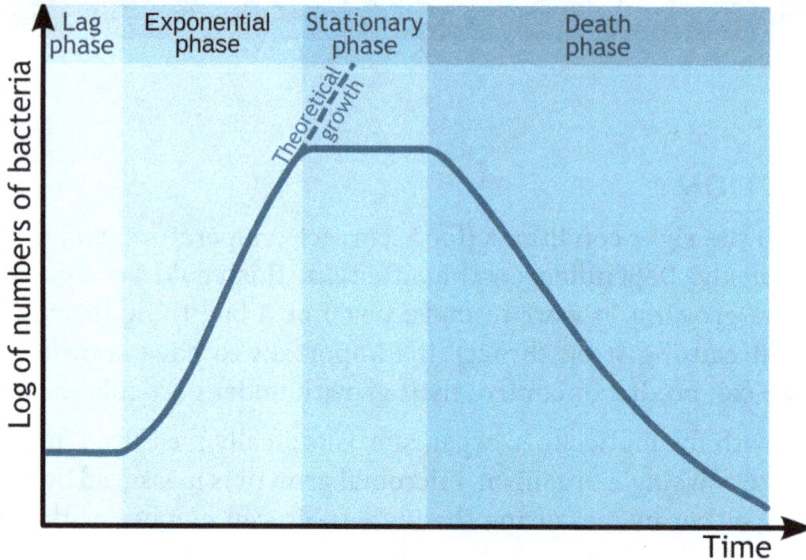

Bacterial Growth Curve. By Michał Komorniczak. If you use on your website or in your publication my images (either original or modified), you are requested to give me details: Michał Komorniczak (Poland) or Michal Komorniczak (Poland).

The details associated with each growth curve (number of cells, length of each phase, rapidness of growth or death, overall amount of time) will vary from organism to organism or even with different conditions for the same organism. But the pattern of four distinct phases of growth will typically remain.

LAG PHASE

The lag phase is an adaptation period, where the bacteria are adjusting to their new conditions. The length of the lag phase can vary considerably, based on how different the conditions are from the conditions that the bacteria came from, as well as the condition of the bacterial cells themselves. Actively growing cells transferred from one type of media into the same type of media, with the same environmental conditions, will have the shortest lag period. Damaged cells will have a long lag period, since they must repair themselves before they can engage in reproduction.

Typically cells in the lag period are synthesizing RNA, enzymes, and essential metabolites that might be missing from their new environment (such as growth factors or macromolecules), as well as adjusting to environmental changes such as changes in temperature, pH, or oxygen availability. They can also be undertaking any necessary repair of injured cells.

Exponential or Log phase

Once cells have accumulated all that they need for growth, they proceed into cell division. The exponential or log phase of growth is marked by predictable doublings of the population, where 1 cell become 2 cells, becomes 4, becomes 8 etc. Conditions that are optimal for the cells will result in very rapid growth (and a steeper slope on the growth curve), while less than ideal conditions will result in slower growth. Cells in the exponential phase of growth are the healthiest and most uniform, which explains why most experiments utilize cells from this phase.

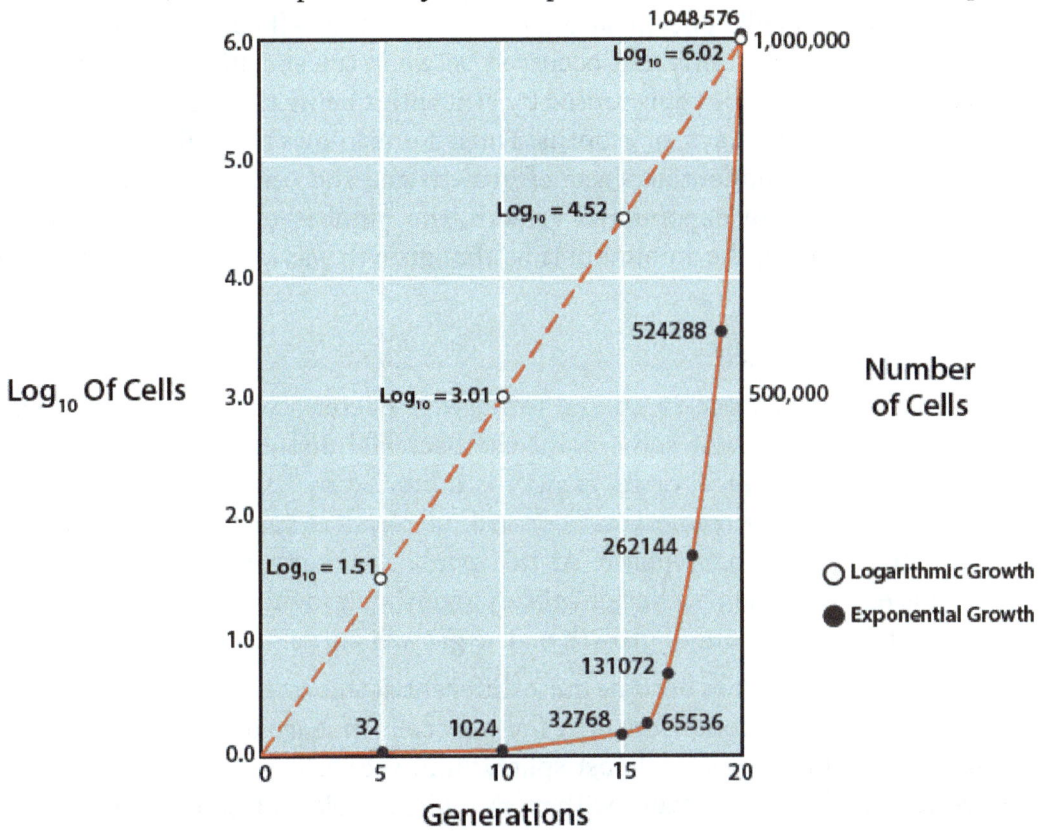

Bacterial Growth Rates.

Since the relationship between time and number of cells is exponential instead of linear, plotting the cell concentration on a semilog scale will standardize the data, giving the appearance of a linear relationship.

Due to the predictability of growth in this phase, this phase can be used to mathe-matically calculate the time it takes for the bacterial population to double in num-ber, known as the generation time (g). This information is used by microbiolo-gists in basic research, as well as in industry. In order to determine generation time, the natural logarithm of cell number can be plotted against time (where the units can vary, depending upon speed of growth for the particular popula-tion), using a semilogarithmic graph to generate a line with a predictable slope. Alternatively one can rely on the fixed relationship between the initial number of cells at the start of the exponential phase and the number of cells after some period of time, which can be expressed by:

$$N=N_0{}^2nN=N_0 2^n$$

where N is the final cell concentration, N0 is the initial cell concentration, and n is the number of generations that occurred between the specified period of time. Generation time (g) can be represented by t/n, with t being the specified period of time in minutes, hours, days, or months. Thus, if one knows the cell concentration at the start of the exponential phase of growth and the cell concentration after some period of time of exponential growth, the number of generations can be calculated. Then, using the amount of time that growth was allowed to proceed (t), one can calculate g.

STATIONARY PHASE

All good things must come to an end (otherwise bacteria would equal the mass of the Earth in 7 days!). At some point the bacterial population runs out of an essential nutrient/chemical or its growth is inhibited by its own waste products (it is a closed container, remember?) or lack of physical space, causing the cells to enter into the stationary phase. At this point the number of new cells being produced is equal to the number of cells dying off or growth has entirely ceased, resulting in a flattening out of growth on the growth curve.

Physiologically the cells become quite different at this stage, as they try to adapt to their new starvation conditions. The few new cells that are produced are smaller in size, with bacilli becoming almost spherical in shape. Their plasma membrane becomes less fluid and permeable, with more hydrophobic molecules on the surface that promote cell adhesion and aggregation. The nucleoid condenses and the DNA becomes bound with DNA-binding proteins from starved cells (DPS), to protect the DNA from damage. The changes are designed to allow the cell to survive for a longer period of time in adverse conditions, while waiting for more optimal conditions (such as an infusion of nutrients) to occur. These same strategies are used by cells in oligotrophic or low-nutrient environments. It has been hypothesized that cells in the natural world (i.e. outside of the laboratory) typically exist for long periods of time in oligotrophic environments, with only sporadic infusions of nutrients that return them to exponential growth for very brief periods of time.

During the stationary phase cells are also prone to producing secondary metabolites, or metabolites produced after active growth, such as antibiotics. Cells that are capable of making an endospore will activate the necessary genes during this stage, in order to initiate the sporulation process.

DEATH OR DECLINE PHASE

In the last phase of the growth curve, the death or decline phase, the number of viable cells decreases in a predictable (or exponential) fashion. The steepness of the slope corresponds to how fast cells are losing viability. It is thought that the culture conditions have deteriorated to a point where the cells are irreparably harmed, since cells collected from this phase fail to show growth when transferred to fresh medium. It is important to note that if the turbidity of a culture is being measured as a way to determine cell density, measurements might not decrease during this phase, since cells could still be intact.

It has been suggested that the cells thought to be dead might be revived under specific conditions, a condition described as viable but nonculturable (VBNC). This state might be of importance for pathogens, where they enter a state of very low metabolism and lack of cellular division, only to resume growth at a later time, when conditions improve.

It has also been shown that 100% cell death is unlikely, for any cell population, as the cells mutate to adapt to their environmental conditions, however harsh. Often there is a tailing effect observed, where a small population of the cells cannot be killed off. In addition, these cells might benefit from their death of their fellow cells, which provide nutrients to the environment as they lyse and release their cellular contents.

KEY WORDS

Binary fission, multiple fission, budding, spores, cell cycle, closed system, batch culture, growth curve, lag phase, exponential or log phase, generation time (g), N, N0, n, t, stationary phase, DNA-binding proteins from starved cells (DPS), oligotrophic, secondary metabolites, death or decline phase, viable but nonculturable (VBNC).

10 | Environmental Factors

INTRODUCTION: AN OVERVIEW

What do real estate agents always say? Location, location, location! It's all about where you live, or at least adapting to where you live. At least it is for microbes.

Competition is fierce out in the microbial world (non-microbial world, too!) and resources can be scarce. For those microbes that are willing and able to adapt to what might be considered a harsh environment, it can certainly mean less competition.

So what environmental conditions can affect microbial growth? Temperature, oxygen, pH, water activity, pressure, radiation, lack of nutrients...these are the primary ones. We will cover more about metabolism (i.e. what type of food can they eat?) later, so let us focus now on the physical characteristics of the environment and the adaptations of microbes.

OSMOLARITY

Cells are subject to changes in osmotic pressure, due to the fact that the plasma membrane is freely permeable to water (a process known as passive diffusion). Water will generally move in the direction necessary to try and equilibrate the cell's solute concentration to the solute concentration of the surrounding environment. If the solute concentration of the environment is lower than the solute concentration found inside the cell, the environment is said to be hypotonic. In this situation water will pass into the cell, causing the cell to swell and increasing internal pressure. If the situation is not rectified then the cell will eventually burst from lysis of the plasma membrane. Conversely, if the solute concentration of the environment is higher than the solute concentration found inside the cell, the environment is said to be hypertonic. In this situation water will leave the cell, causing the cell to dehydrate. Extended periods of dehydration will cause permanent damage to the plasma membrane.

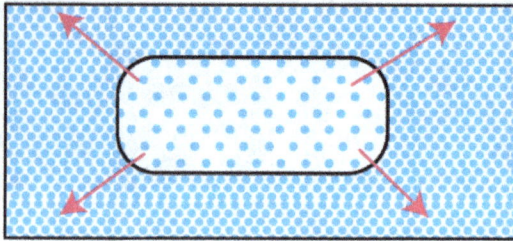

Hypertonic Solution   Hypotonic Solution

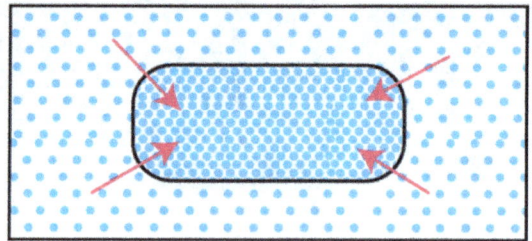

Hypertonic vs. Hypotonic Solutions.

Cells in a hypotonic solution need to reduce the osmotic concentration of their cytoplasm. Sometimes cells can use inclusions to chemically change their solutes, reducing molarity. In a real pinch they can utilize what are known as mechanosensitive (MS) channels, located in their plasma membrane. MS channels open as the plasma membrane stretches due to the increased pressure, allowing solutes to leave the cell and thus lowering the osmotic pressure.

Cells in a hypertonic solution needing to increase the osmotic concentration of their cytoplasm can take up additional solutes from the environment. However, cells have to be careful about what solutes they take up, since some solutes can interfere with cellular function and metabolism. Cells need to take up compatible solutes, such as sugars or amino acids, which typically will not interfere with cellular processes.

There are some microbes that have evolved to extreme hypertonic environments, specifically high salt concentrations, to the point where they now require the presence of high levels of sodium chloride to grow. Halophiles, which require a NaCl concentration above 0.2 M, take in both potassium and chloride ions as a way to offset the effects of the hypertonic environment that they live in. Their evolution has been so complete that their cellular components (ribosomes, enzymes, transport proteins, cell wall, plasma membrane) now require the presence of high concentrations of both potassium and chloride to function.

pH

pH is defined as the negative logarithm of the hydrogen ion concentration of a solution, expressed in molarity. The pH scale ranges from 0 to 14, with 0 representing an extremely acidic solution (1.0 M H+) and 14 representing an extremely alkaline solution (1.0 x 14-10 M H+). Each pH units represents a tenfold change in hydrogen ion concentration, meaning a solution with a pH of 3 is 10x more acidic than a solution with a pH of 4.

Typically cells would prefer a pH that is similar to their internal environment, with cytoplasm having a pH of 7.2. That means that most microbes are neutrophiles ("neutral lovers"), preferring a pH in the range of 5.5 to 8.0. There are some microbes, however, that have evolved to live in the extreme pH environments.

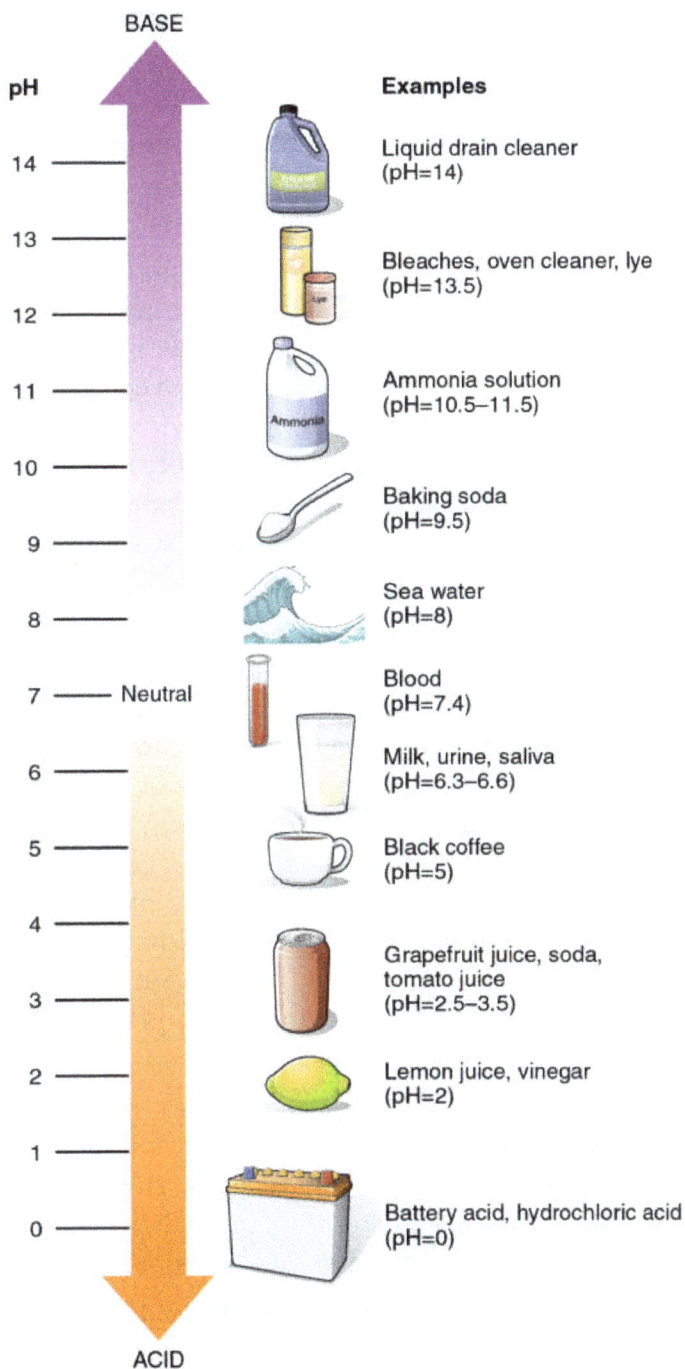

pH Scale. OpenStax, Inorganic Compounds Essential to Human Functioning.
OpenStax CNX. Jun 18, 2013 Temperature

Acidophiles ("acid lovers"), preferring an environmental pH range of 0 to 5.5, must use a variety of mechanisms to maintain their internal pH in an acceptable range and preserve the stability of their plasma membrane. These organisms transport cations (such as potassium ions) into the cell, thus decreasing H+ movement into the cell. They also utilize proton pumps that actively pump H+ out.

Alkaliphiles ("alkaline lovers"), preferring an environmental pH range of 8.0 to 11.5, must pump protons in, in order to maintain the pH of their cytoplasm. They typically employ antiporters, which pump protons in and sodium ions out.

Microbes have no way to regulate their internal temperature so they must evolve adaptations for the environment they would like to live in. Changes in temperature have the biggest effect on enzymes and their activity, with an optimal temperature that leads to the fastest metabolism and resulting growth rate. Temperatures below optimal will lead to a decrease in enzyme activity and slower metabolism, while higher temperatures can actually denature proteins such as enzymes and carrier proteins, leading to cell death. As a result, microbes have a growth curve in relation to temperature with an optimal temperature at which growth rate peaks, as well as minimum and maximum temperatures where growth continues but is not as robust. For a bacterium the growth range is typically around 30 degrees.

The psychrophiles are the cold lovers, with an optimum of 15°C or lower and a growth range of -20°C to 20°C. Most of these microbes are found in the oceans, where the temperature is often 5°C or colder. They can also be found in the Arctic and the Antarctic, living in ice wherever they can find pockets of liquid water. Adaptation to the cold required evolution of specific proteins, particularly enzymes, that can still function in low temperatures. In addition, it also required modification to the plasma membrane to keep it semifluid. Psychrophiles have an increased amount of unsaturated and shorter-chain fatty acids. Lastly, psychrophiles produce cryoprotectants, special proteins or sugars that prevent the development of ice crystals that might damage the cell. Psychrotophs or cold tolerant microbes have a range of 0-35°C, with an optimum of 16°C or higher.

Humans are best acquainted with the mesophiles, microbes with a growth optima of 37°C and a range of 20-45°C. Almost all of the human microflora fall into this category, as well as almost all human pathogens. The mesophiles occupy the same environments that humans do, in terms of foods that we eat, surfaces that we touch, and water that we drink and swim in.

On the warmer end of the spectrum is where we find the thermophiles ("heat lovers"), the microbes that like high temperatures. Thermophiles typically have a range of 80-45°C, and a growth optimum of 60°C. There are also the hyperthermophiles, those microbes that like things extra spicy. These microbes have a growth optima of 88-106°C, a minimum of 65°C and a maximum of 120°C.

Both the thermophiles and the hyperthermophiles require specialized heat-stable enzymes that are resistant to denaturation and unfolding, partly due to the presence of protective proteins known as chaperone proteins. The plasma membrane of these organisms contains more saturated fatty acids, with increased melting points.

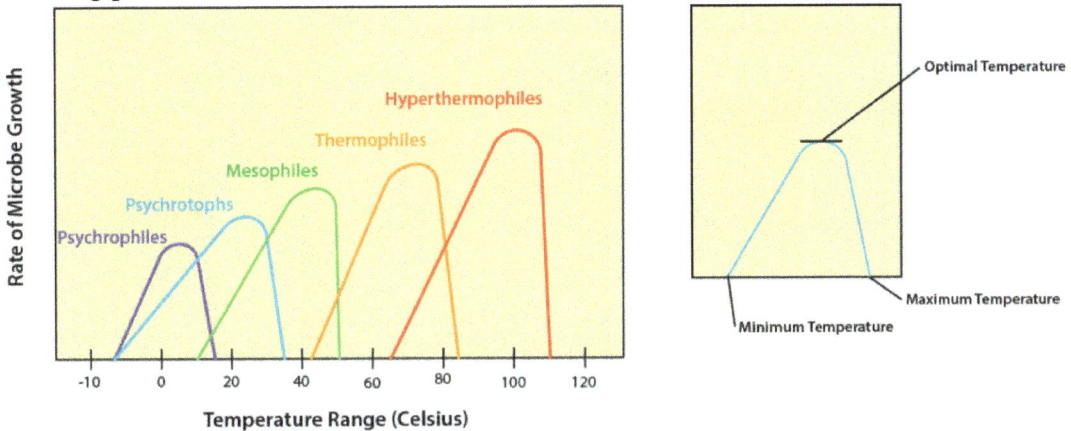

Growth Curves.

OXYGEN CONCENTRATION

The oxygen requirement of an organism relates to the type of metabolism that it is using. Energy generation is tied to the movement of electrons through the electron transport chain (ETC), where the final electron acceptor can be oxygen or a non-oxygen molecule.

Organisms that use oxygen as the final electron acceptor are engaging in aerobic respiration for their metabolism. If they require the presence of atmospheric oxygen (%20) for their metabolism then they are referred to as obligate aerobes. Microaerophiles require oxygen, but at a lower level than normal atmospheric levels – they only grow at levels of %10-2.

Organisms that can grow in the absence of oxygen are referred to as anaerobes, with several different categories existing. The facultative anaerobes are the most versatile, being able to grow in the presence or absence of oxygen by switching their metabolism to match their environment. They would prefer to grow in the presence of oxygen, however, since aerobic respiration generates the largest amount of energy and allows for faster growth. Aerotolerant anaerobes can also grow in the presence or absence of oxygen, exhibiting no preference. Obligate anaerobes can only grow in the absence of oxygen and find an oxygenated environment to be toxic.

While the use of oxygen is dictated by the organism's metabolism, the ability to live in an oxygenated environment (regardless of whether it is used by the organism or not) is dictated by the presence/absence of several enzymes. Oxygen by-products (called reactive oxygen species or ROS) can be toxic to cells, even to the cells using aerobic respiration. Enzymes that can offer some protection from ROS include superoxide dismutase (SOD), which breaks down superoxide radicals, and catalase, which breaks down hydrogen peroxide. Obligate anaerobes lack both enzymes, leaving them little or no protection against ROS.

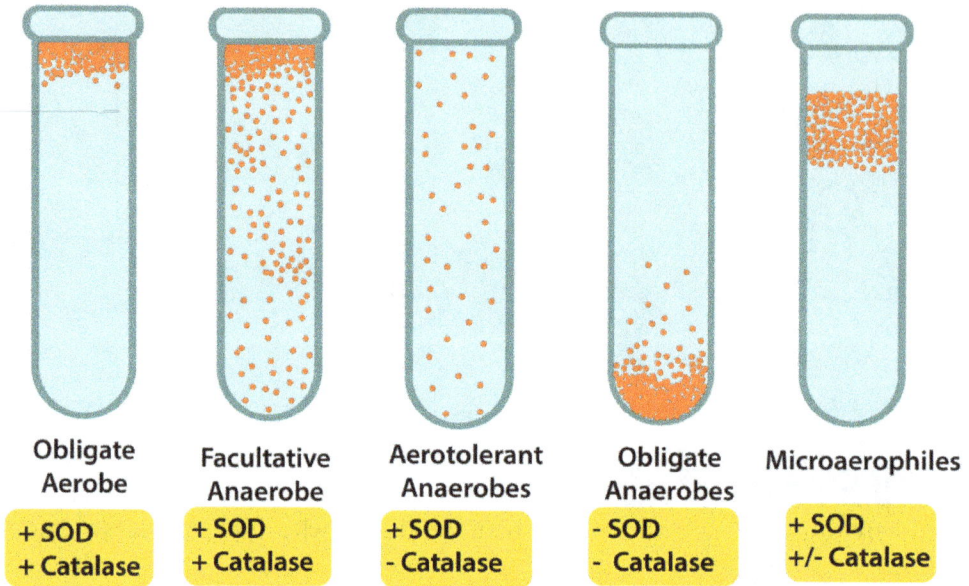

Obligate Aerobe	Facultative Anaerobe	Aerotolerant Anaerobes	Obligate Anaerobes	Microaerophiles
+ SOD + Catalase	+ SOD + Catalase	+ SOD - Catalase	- SOD - Catalase	+ SOD +/- Catalase

Oxygen and Bacterial Growth.

PRESSURE

The vast majority of microbes, living on land or water surface, are exposed to a pressure of approximately 1 atmosphere. But there are microbes that live on the bottom of the ocean, where the hydrostatic pressure can reach 600-1,000 atm. These microbes are the barophiles ("pressure lovers"), microbes that have adapted to prefer and even require the high pressures. These microbes have increased unsaturated fatty acids in their plasma membrane, as well as shortened fatty acid tails.

RADIATION

All cells are susceptible to the damage cause by radiation, which adversely affects DNA with its short wavelength and high energy. Ionizing radiation, such as x-rays and gamma rays, causes mutations and destruction of the cell's DNA. While bacterial endospores are extremely resistant to the harmful effects of ionizing radiation, vegetative cells were thought to be quite susceptible to its impact. That is, until the discovery of Deinococcus radiodurans, a bacterium capable of completely reassembling its DNA after exposure to massive doses of radiation.

Ultraviolet (UV) radiation also causes damage to DNA, by attaching thymine bases that are next to one another on the DNA strand. These thymine dimers inhibit DNA replication and transcription. Microbes typically have DNA repair mechanisms that allow them to repair limited damage, such as the enzyme photolyase that splits apart thymine dimers.

Key Words

Osmotic pressure, passive diffusion, solute, hypotonic, hypertonic, mechanosensitive (MS) channel, compatible solute, halophile, pH, neutrophile, acidophile, alkaliphile, optimum temperature, minimum temperature, maximum temperature, psychrophile, psychrotroph, mesophile, thermophile, hyperthermophile, chaperone protein, electron transport chain (ETC), aerobic respiration, obligate aerobe, microaerophile, anaerobe, facultative anaerobe, aerotolerant anaerobe, obligate anaerobe, reactive oxygen species (ROS), superoxide dismutase (SOD), catalase, barophile, ionizing radiation, Deinococcus radiodurans, ultraviolet (UV) radiation, thymine dimmers, photolyase.

NUTRITIONAL TYPES OF MICROORGANISMS

All microbes have a need for three things: carbon, energy, and electrons. There are specific terms associated with the source of each of these items, to help define organisms.

Let us focus on carbon first. All organisms are carbon-based with macromolecules proteins, carbohydrates, lipids, nucleic acid – having a fundamental core of carbon. On one hand, organisms can use reduced, preformed organic substances as a carbon source. These are the heterotrophs or "other eaters." Alternatively, they can rely on carbon dioxide (CO_2) as a carbon source, reducing or "fixing" it this inorganic form of carbon into an organic molecule. These are the autotrophs or "self feeders."

For energy, there are two possibilities as well: light energy or chemical energy. Light energy comes from the sun, while chemical energy can come from either organic or inorganic chemicals. Those organisms that use light energy are called phototrophs ("light eaters"), while those that use chemical energy are called chemotrophs ("chemical eaters"). Chemical energy can come from inorganic sources or organic sources. An organism that uses inorganic sources is known as a lithotroph ("rock eater"), while an organism that uses organic sources is called an organotroph ("organic eater").

These terms can all be combined, to derive a single term that gives you an idea of what an organism is using to meet its basic needs for energy, electrons, and carbon.

Nutritional Type of Microorganisms

Energy Source	Electron Source	Carbon Source	Nutritional Type
Light Photo	Organic -Organo-	Organic -heterotroph-	Photooganoheterotroph
		Carbon dioxide -hererotroph-	(no know organisms)
	Inorganic -litho-	Organic -heterotroph-	(no know organisms)

Energy Source	Electron Source	Carbon Source	Nutritional Type
		Carbon dioxide -hererotroph-	Photolithoautotroph
Chemical Compounds Chemo-	Organic -Organo-	Organic -heterotroph-	Chemoorganoheterotroph
		Carbon dioxide -hererotroph-	(no know organisms)
	Inorganic -litho-	Organic -heterotroph-	Chemolithohctcrotroph
		Carbon dioxide -hererotroph-	Chemolithoautotroph

MACRONUTRIENTS

In addition to carbon, hydrogen and oxygen, cells need a few other elements in sufficient quantity. In particular, cells need nitrogen for the formation of proteins, nucleic acids, and a few other cell components. Cells also need phosphorous, which is a crucial component of nucleic acids (think sugar-phosphate backbone!), phospholipids, and adenosine triphosphate or ATP. Sulfur is necessary for a few amino acids, as well as several vitamins, while potassium is needed for enzymes, and magnesium is used to stabilize ribosomes and membrane. Collectively these elements (including C, H, and O) are referred to as the macronutrients.

GROWTH FACTORS

Some microbes can synthesize certain organic molecules that they need from scratch, as long as they are provided with carbon source and inorganic salts. Other microbes require that certain organic compounds exist within their environment. These organic molecules essential for growth are called growth factors and fall in three categories: 1) amino acids (building blocks of protein), 2) purines and pyrimidines (building blocks of nucleic acid), and 3) vitamins (enzyme cofactors).

UPTAKE OF NUTRIENTS

In order to support its' activities, a cell must bring in nutrients from the external environment across the cell membrane. In bacteria and archaea, several different transport mechanisms exist.

PASSIVE DIFFUSION

Passive or simple diffusion allows for the passage across the cell membrane of simple molecules and gases, such as CO_2, O_2, and H_2O. In this case, a concentration gradient must exist, where there is higher concentration of the substance outside of the cell than there is inside the cell. As more of the substance is transported into the cell the concentration gradient decreases, slowing the rate of diffusion.

FACILITATED DIFFUSION

Facilitated diffusion also involves the use of a concentration gradient, where the concentration of the substance is higher outside the cell, but differs with the use of carrier proteins (sometimes called permeases). These proteins are embedded within the cell membrane and provide a channel or pore across the membrane barrier, allowing for the passage of larger molecules. If the concentration gradient dissipates, the passage of molecules into the cell stops. Each carrier protein typically exhibits specificity, only transporting in a particular type of molecule or closely related molecules.

ACTIVE TRANSPORT

Many types of nutrient uptake require that a cell be able to transport substances against a concentration gradient (i.e. with a higher concentration inside the cell than outside). In order to do this, a cell must utilize metabolic energy for the transport of the substance through carrier proteins embedded in the membrane. This is known as active transport. All types of active transport utilize carrier proteins.

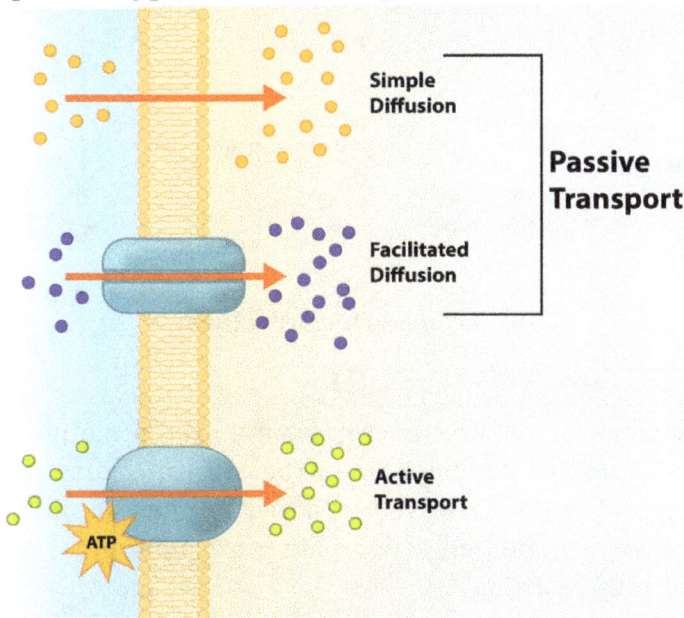

Active Transport Versus Passive Transport

PRIMARY ACTIVE TRANSPORT

Primary active transport involves the use of chemical energy, such as ATP, to drive the transport. One example is the ABC system, which utilizes ATP-Binding Cassette transporters. Each ABC transporter is composed of three different components: 1) membrane-spanning proteins that form a pore across the cell membrane (i.e. carrier protein), 2) an ATP binding region that hydrolyzes ATP, providing the energy for the passage across the membrane, and 3) a substrate-binding protein, a peripheral protein that binds to the appropriate substance to be transporter and ferries it to the membrane-spanning proteins. In gram negative bacteria the substrate-binding protein is located in the cell's periplasm, while in gram positive bacteria the substrate-binding protein is attached to the outside of the cell membrane.

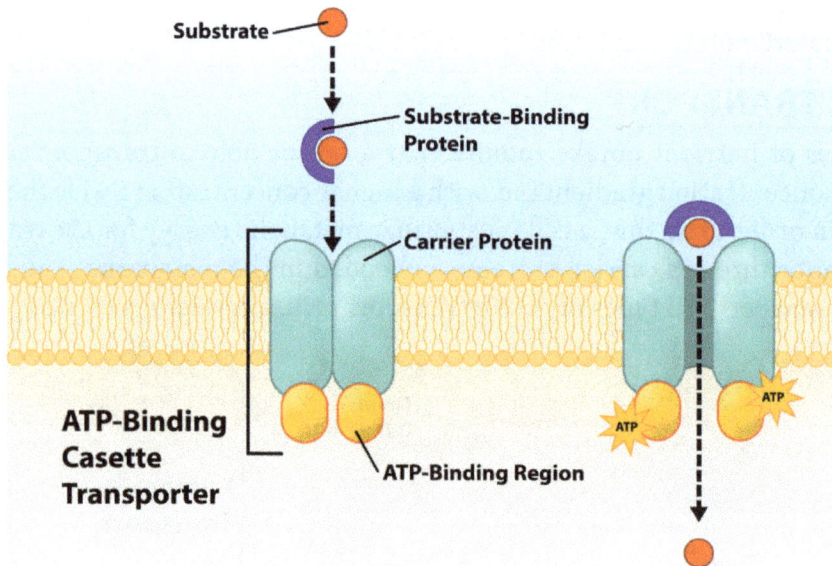

ABC Transporter Structure.

SECONDARY ACTIVE TRANSPORT

Secondary active transport utilizes energy from a proton motive force (PMF). A PMF is an ion gradient that develops when the cell transports electrons during energy-conserving processes. Positively charged protons accumulate along the outside of the negatively charged cell, creating a proton gradient between the outside of the cell and the inside.

There are three different types of transport events for simple transport: uniport, symport, and antiport and each mechanism utilizes a different protein porter. Uniporters transport a single substance across the membrane, either in or out. Symporters transport two substances across the membrane at the same time, typically a proton paired with another molecule. Antiporters transport two substances across the membrane as well, but in opposite directions. As one substance enters the cell, the other substance is transported out.

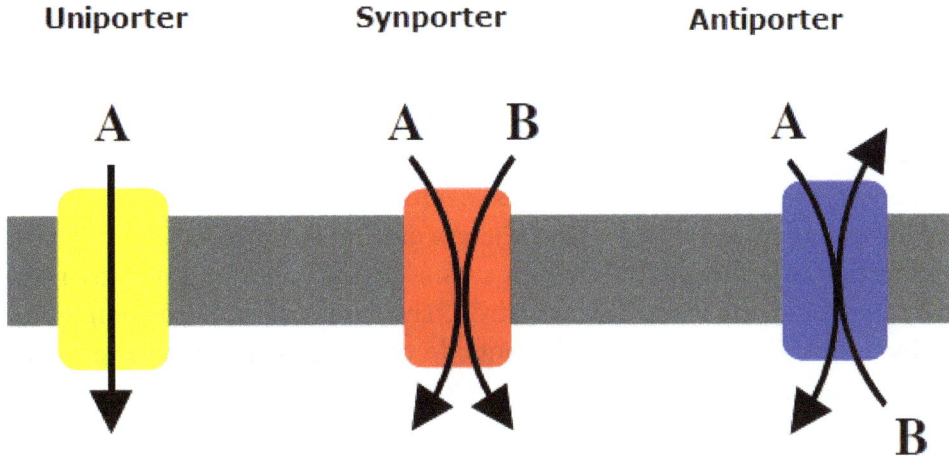

Uniporter   Synporter   Antiporter

Uniport Synport Antiport. By Lupask (Own work) [Public domain], via Wikimedia Commons

GROUP TRANSLOCATION

Group translocation is a distinct type of active transport, using energy from an energy-rich organic compound that is not ATP. Group translocation also differs from both simple transport and ABC transporters in that the substance being transported is chemically modified in the process.

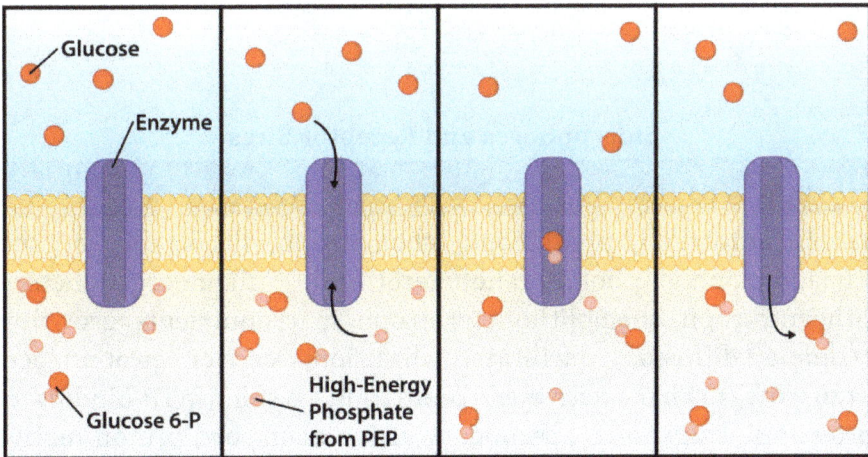

Group Translocation via PTS.

One of the best studied examples of group translocation is the phosphoenol-pyruvate: sugar phosphotransferase system (PTS), which uses energy from the high-energy molecule phosphoenolpyruvate (PEP) to transport sugars into the cell. A phosphate is transferred from the PEP to the incoming sugar during the process of transportation.

IRON UPTAKE

Iron is required by microbes for the function of their cytochromes and enzymes, resulting in it being a growth-limiting micronutrient. However, little free iron is available in environments, due to its insolubility. Many bacteria have evolved siderophores, organic molecules that chelate or bind ferric iron with high affinity. Siderophores are released by the organism to the surrounding environment, whereby they bind any available ferric iron. The iron-siderophore complex is then bound by a specific receptor on the outside of the cell, allowing the iron to be transported into the cell.

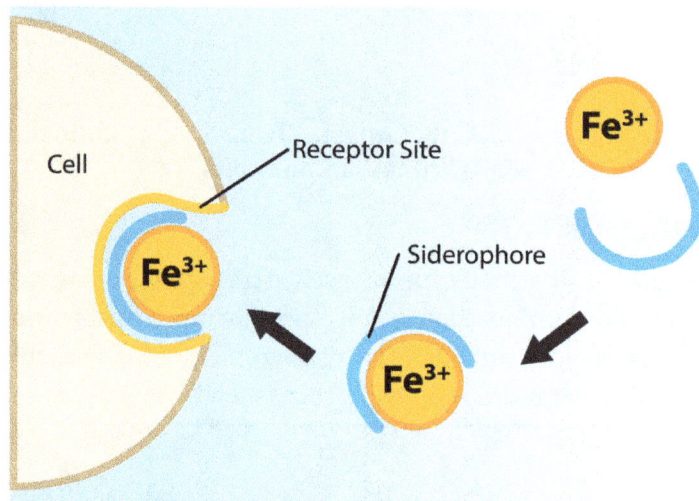

Siderophores and Receptor Sites.

KEY WORDS

Heterotroph, autotroph, phototroph, chemotroph, lithotroph, organotroph, photolithoautotroph, photoorganoheterotroph, chemoorganoheterotroph, chemolithoautotroph, chemolithoheterotroph, macronutrients, growth factors, passive/simple diffusion, facilitated diffusion, carrier protein/permease, active transport, primary active transport, ABC system, ATP-binding cassette transporter, ABC transporter, secondary active transport, proton motive force (PMF), uniport, symport, antiport, porter, uniporter, symporter, antiporter, group translocation, phosphoenolpyruvate: sugar phosphotransferase system (PTS), phosphoenolpyruvate (PEP), siderophore.

12 | Energetics & Redox Reactions

METABOLISM

Metabolism refers to the sum of chemical reactions that occur within a cell. Catabolism is the breakdown of organic and inorganic molecules, used to release energy and derive molecules that could be used for other reactions. Anabolism is the synthesis of more complex molecules from simpler organic and inorganic molecules, which requires energy.

ENERGETICS

While some energy is lost as heat in chemical reactions, the measurement of interest for cells is the amount of free energy (G), or the energy available to do work. Cells perform three different types of work: chemical work (such as anabolism), transport work (such as nutrient uptake), and mechanical work (such as the rotation of a flagellum).

The change in free energy is typically denoted as $\Delta G°'$, which indicates the change in free energy under standard conditions of pH 7, 25oC, 1 atmosphere pressure (also known as the standard free energy change). A reaction that generates a positive $\Delta G°'$ indicates that the reaction requires energy and is endergonic in nature. A reaction that generates a negative $\Delta G°'$ indicates that the reaction releases energy and is exergonic in nature. Reactions that are exergonic release energy that can be conserved by the cell to do work.

ADENOSINE TRIPHOSPHATE (ATP)

Adenosine triphosphate or ATP is a high-energy molecule used by all cells for energy currency, partly because it readily donates a phosphoryl group to other molecules. An exergonic reaction will release energy, driving the synthesis of ATP from the addition of a phosphate molecule (orthophosphate or Pi) to adenosine diphosphate or ADP. An endergonic reaction, which requires energy, will couple with the hydrolysis of ATP to ADP + Pi, using the released energy to drive the reaction.

ENZYMES

In order for a chemical reaction to proceed, chemical bonds must be broken.

The energy required to break bonds is called activation energy. The amount of activation energy required by a cell can be lowered with the help of a catalyst, substances which assist the reaction to proceed without being changed themselves by the reaction. Cells use protein catalysts known as enzymes.

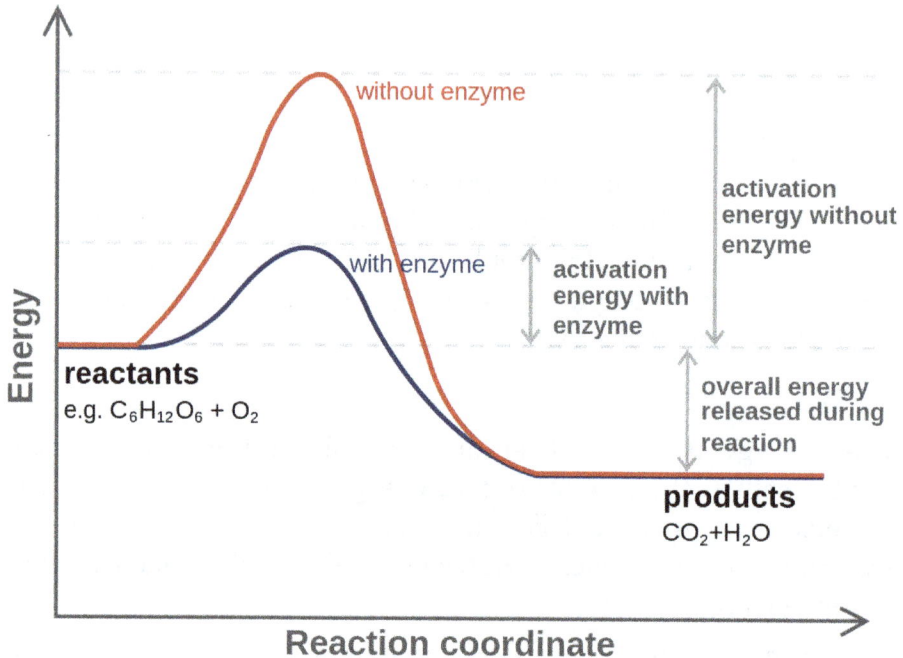

Activation energy. By Originally uploaded by Jerry Crimson Mann, vectorized by Tutmosis, corrected by Fvasconcellos (en:Image:Activation2.png) [GFDL or CC-BY-SA-3.0], via Wikimedia Commons

REDOX REACTIONS

Cells conserve energy in the form of ATP by coupling its synthesis to the release of energy via oxidation-reduction (redox) reactions, where electrons are passed from an electron donor to an electron acceptor. The oxidation of a molecule refers to the loss of its electrons, while the reduction of a molecule refers to its gain of electrons. Organic chemists often refer to the process by the mnemonic OIL RIG: Oxidation Is Loss, Reduction Is Gain. A molecule being oxidized is acting as an electron donor, while the molecule being reduced is acting as an electron acceptor. Since electrons represent energy, a substance with many electrons to donate can be thought of as energy-rich.

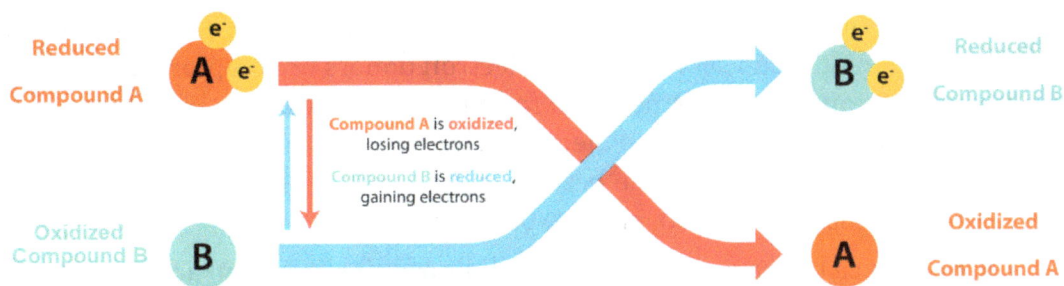

Conjugate Redox Pair

Electrons do not exist freely in solution, they must be coupled with atoms or molecules. Every redox reaction consists of two half reaction, where one substance donates electrons and thus becomes an oxidized product while another substance accepts the electrons and thus becomes a reduced product. Conjugate redox pair refers to the acceptor and donor of a half reaction.

Every redox reaction consists of two half reactions, where the reduced substance can donate electrons and thus become an oxidized product, while the oxidized substance can accept electrons and thus become a reduced product in a different reaction. A substance can be either an electron donor or an electron acceptor, dependent upon the other substances in the reaction.

An example would be ½ O2/H_2O, where H_2O could serve as an electron donor in one reaction, becoming the product O_2 as a result of being oxidized, while O_2 could serve as an electron acceptor in a different reaction, becoming the product H_2O as a result of being reduced. In order for either reaction to be complete, another redox couple would need to participate

Each half reaction is given a standard reduction potential (E'0) in volts or millivolts, which is a measurement of the tendency of the donor in the reaction to give up electrons. A substance with greater tendency to donate electrons in the reduced form has a more negative E'0, while a substance with a weak tendency to donate electrons in the reduced form has a less negative or even positive E'0. A substance with a negative E'0 makes a very good electron donor, in the reduced form.

REDOX TOWER

The information regarding standard reduction potentials for various redox couples is displayed in the form of a redox tower, which lists the couples in a vertical form based on their E'0. Redox couples with the most negative E'0 on listed at the top while those with the most positive E'0 are listed on the bottom. The reduced substance with the greatest tendency to donate electrons would be found at the top of the tower on the right, while the oxidized substance with the greatest tendency to accept electrons would be found at the bottom of the tower on the left. Redox couples in the middle can serve as either electron donors or acceptors, depending upon what substance they partner with for a reaction. Applying this information to the redox couples shown in the electron tower below, the best initial electron donor would be glucose, while the best electron acceptor would be oxygen. This would result in the products CO_2 and H_2O being formed.

The difference between reduction potentials of a donor and an acceptor ($\Delta E'0$) is measured as acceptor E'0 minus donor E'0. The larger the value for $\Delta E'0$, the more potential energy for a cell. Larger values are derived when there is the biggest distance between the donor and the acceptor (or a bigger fall down the tower).

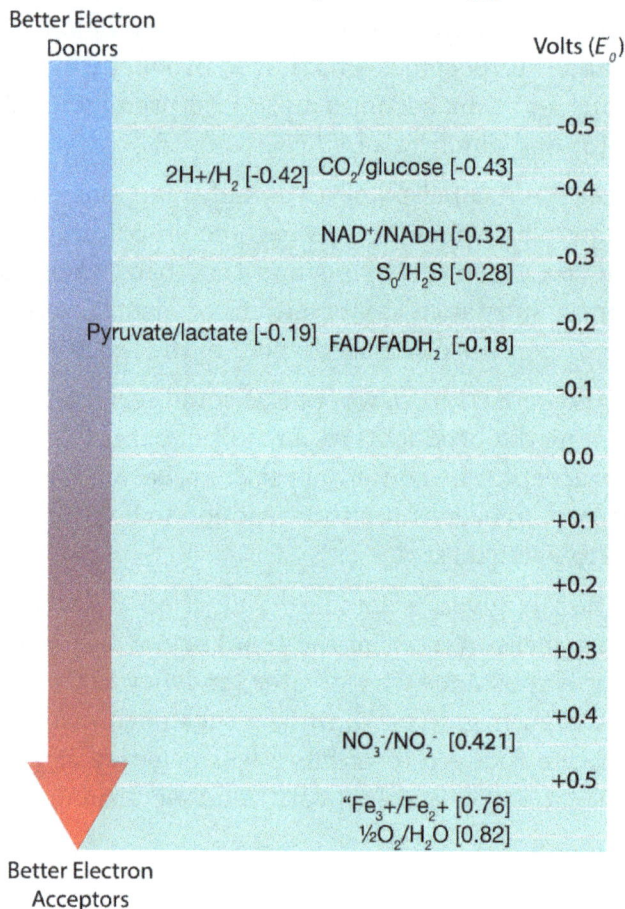

Better Electron
Donors   Volts (E_0)

-0.5

$2H+/H_2$ [-0.42] CO_2/glucose [-0.43]
-0.4

NAD$^+$/NADH [-0.32]
-0.3
S_0/H_2S [-0.28]

Pyruvate/lactate [-0.19] FAD/FADH$_2$ [-0.18]
-0.2

-0.1

0.0

+0.1

+0.2

+0.3

+0.4

NO$_3^-$/NO$_2^-$ [0.421]
+0.5

"Fe$_3$+/Fe$_2$+ [0.76]
½O$_2$/H$_2$O [0.82]

Better Electron
Acceptors

Electron Tower.

While $\Delta E'0$ is proportional to $\Delta G°'$, the number of electrons that a substance has to donate is important too. The actual formula is:

$$\Delta G°'=-nF \cdot \Delta E'0 \Delta G°'=-nF \cdot \Delta E'0$$

Where n is the number of electrons being transferred and F is the Faraday constant (23,062 cal/mole-volt, 96, 480 J/mole-volt).

ELECTRON CARRIERS

The transference of electrons from donor to acceptor does not occur directly, since chemically dissimilar electron donors and acceptors might never interact with one another. Instead, many cellular intermediates participate in the process, with the possibility for energy capture occurring along the way. These intermediates are called electron carriers and they go back and forth between a reduced form (when they are carrying an electron) and an oxidized form (after they have passed the electron on), without being consumed in the reaction themselves. It is important to note that electron carriers can never serve as the initial electron donor or the final electron acceptor for a reaction, since they originate within the cell itself and need to be constantly recycled in order to continually participate in reactions. The cell needs to rely upon external chemicals that it transports in to act as the initial electron donor and the final electron acceptor. Products that are formed as a result of the reaction (i.e. the now oxidized electron donor and the now reduced electron acceptor) are often waste products for the cell and can be released to the environment.

In order for the reaction to be energetically favorable for the cell, the carriers must be arranged in order of their standard reduction potential (i.e. going down the redox tower), with an electron being passed from a carrier with the most negative E'0 to a carrier with a less negative E'0. It is important to note that some carriers accept both electrons and protons, while other carriers accept electrons only. This fact will become of crucial importance later, in the discussion of how energy is generated.

While there are many different electron carriers, some unique to specific organisms or groups of organisms, let us cover some of the more common ones:

- Nicotinamide adenine dinucleotide (NAD+/NADH) – a co-enzyme that carriers both electrons (e-) and protons (H+), two of each. A closely related molecule is nicotinamide adenine dinucleotide phosphate (NADP+/ NADPH), which accepts 2 electrons and 1 proton.
- Flavin adenine dinucleotide (FAD/FADH) and flavin mononucleotide (FMN/ FMNH) – carry 2 electrons and 2 protons each. Proteins with these molecules are called flavoproteins.
- Coenzyme Q (CoQ)/ubiquinone – carries 2 electrons and 2 protons.
- Cytochromes – use iron atoms as part of a heme group to carry 1 electron at a time.

⊙ Iron-sulfur (Fe-S) proteins, such as ferredoxin – use iron atoms not part of heme group to carry 1 electron at a time.

ELECTRON TRANSPORT CHAIN

The process starts with an initial electron donor, a substance from outside of the cell, and ends with a final electron acceptor, another substance from outside of the cell. In the middle the electrons are passed from carrier to carrier, as the electrons work their way down the electron tower. In order to make the process more efficient, most of the electron carriers are embedded within a membrane of the cell, in the order that they are arranged on a redox tower. These electron transport chains are found within the cell membrane of bacteria and archaea, and within the mitochondrial membrane of eukaryotes.

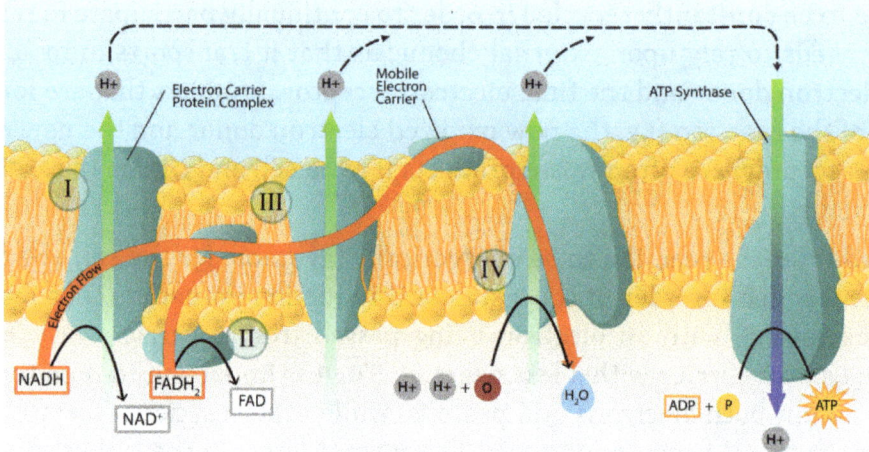

Electron Transport Chain.

KEY WORDS

Metabolism, catabolism, anabolism, free energy (G), chemical work, transport work, mechanical work, $\Delta G°'$, standard free energy change, exergonic, endergonic, adenosine triphosphate (ATP), orthophosphate (Pi), activation energy, catalyst, enzyme, oxidation-reduction (redox) reaction, electron donor, electron acceptor, OIL RIG, conjugate redox pair, redox couple, standard reduction potential (E'0), redox tower, $\Delta E'0$, electron carriers, nicotinamide adenine dinucleotide (NAD+/ NADH), nicotinamide adenine dinucleotide phosphate (NADP+/ NADPH), flavin adenine dinucleotide (FAD/FADH), flavin mononucleotide (FMN/FMNH), coenzyme Q (CoQ)/ubiquinone, cytochrome, iron-sulfur (Fe-S) proteins, ferredoxin, electron transport chain (ETC).

12 | Introduction to Chemoorganotrophy

GENERAL INTRODUCTION

Chemoorganotrophy is a term used to denote the oxidation of organic chemicals to yield energy. In other words, an organic chemical serves as the initial electron donor. The process can be performed in the presence or absence of oxygen, depending upon what is available to a cell and whether or not they have the enzymes to deal with toxic oxygen by-products.

AEROBIC RESPIRATION

To start, let us focus on the catabolism of organic compounds when it occurs in the presence of oxygen. In other words, oxygen is being used as the final electron acceptor. When the process utilizes glycolysis and the tricarboxylic acid (TCA) cycle to completely oxidize an organic compound down to CO_2, it is known as aerobic respiration. This generates the most ATP for a cell, given the large amount of distance between the initial electron donor (glucose) and the final electron acceptor (oxygen), as well as the large number of electrons that glucose has to donate.

ORGANIC ENERGY SOURCES

In chemoorganotrophy, energy is derived from the oxidation of an organic compound. There are many different organic compounds available to a cell, such as proteins, polysaccharides, and lipids. But cellular pathways are arranged in such a way to increase metabolic efficiency. Thus, the cell funnels reactions into a few common pathways. By convention, glucose is used as the starting molecule to describe each process.

GLYCOLYSIS

Glycolysis is a nearly universal pathway for the catabolism of glucose to pyruvate. The pathway is divided into two parts: part I, which focuses on modifications to the 6-carbon sugar glucose, and part II, where the 6-carbon compound is split into two 3-carbon molecules, yielding a bifurcated pathway. Part I actually requires energy in the form of 2 molecules of ATP, in order to phosphorylate or activate the sugar. Part II is the energy conserving phase of the reaction, where 4 molecules of ATP are generated by substrate-level phosphorylation, where a high-energy molecule directly transfers a Pi to ADP.

The net yield of energy from glycolysis is 2 molecules of ATP for every molecule of glucose. In addition, 2 molecules of the carrier NAD+ are reduced, forming NADH. In aerobic respiration, these electrons will ultimately be transferred by NADH to an electron transport chain, allowing the cell to capture more energy. Lastly, 2 molecules of the 3-carbon compound pyruvate are produced, which can be further oxidized to capture more energy for the cell.

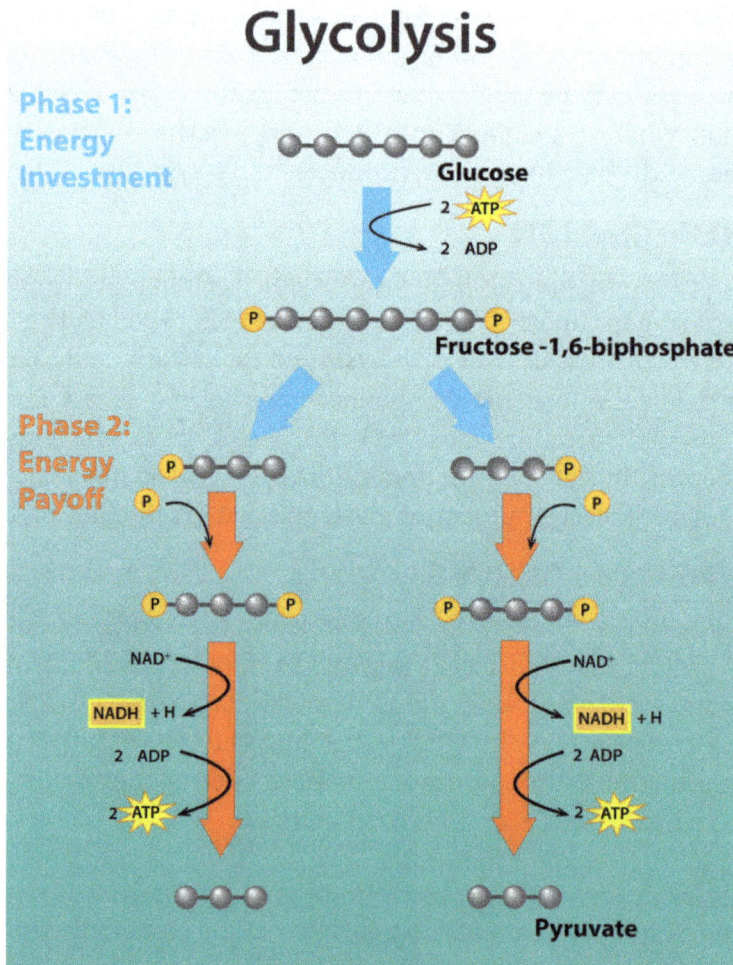

Glycolysis.

TRICARBOXYLIC ACID (TCA) CYCLE

The tricarboxylic acid (TCA) cycle picks up at the end of glycolysis, in order to fully oxidize each molecule of pyruvate down to 3 molecules of CO_2, as occurs in aerobic respiration. It begins with a type of connecting reaction before the molecules can enter the cycle proper. The connecting reaction reduces 1 molecule of NAD+ to NADH for every molecule of pyruvate, in the process of making citrate.

The citrate enters the actual cycle part of the process, undergoing a series of oxidations that yield many different products, many of them important precursor metabolites for other pathways. As electrons are released, carriers are reduced, yielding 3 molecules of NADH and 1 molecule of $FADH_2$ for every molecule of pyruvate. In addition, 1 molecule of GTP (which can be thought of as an ATP-equivalent molecule) is generated by substrate-level phosphorylation.

Taking into account that there were two molecules of pyruvate generated from glycolysis, the net yield of the TCA cycle and its connecting reaction are: 2 molecules of GTP, 8 molecules of NADH, and 2 molecules of $FADH_2$. But where does the ATP come from? So far we only have the net yield of 2 molecules from glycolysis and the 2 molecules of ATP-equivalents (i.e. GTP) from the TCA cycle. This is where the electron transport chain comes into play.

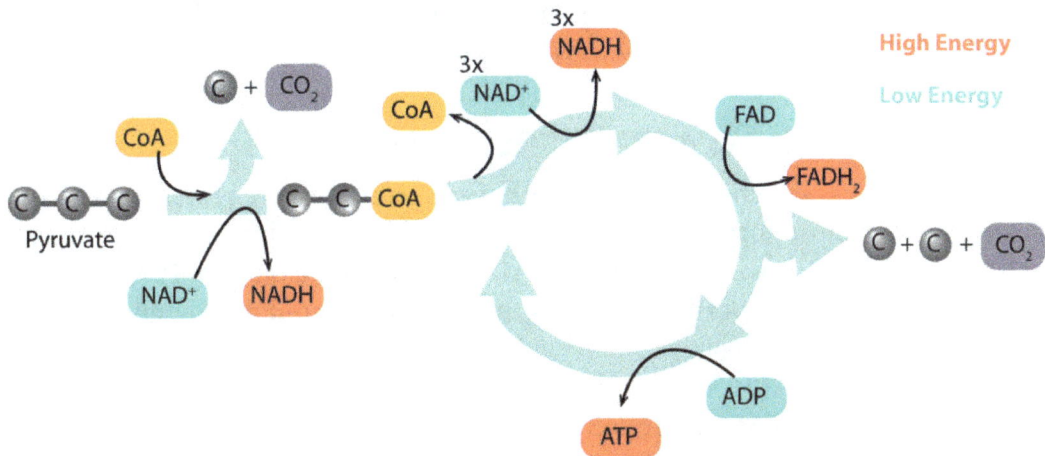

Tricaboxylic Acid Cycle, TCA at the End of Glycolysis.

OXIDATIVE PHOSPHORYLATION

The synthesis of ATP from electron transport generated from oxidizing a chemical energy source is known oxidative phosphorylation. We have already established that electrons get passed from carrier to carrier, in order of their standard reduction potential. We have also established that some carriers accept electrons and protons, while others accept electrons only. What happens to the unaccepted protons? And how does this generate ATP for the cell? Welcome to the wonderful world of the proton motive force (PMF) and ATP synthase!

PROTON MOTIVE FORCE

Protons that are not accepted by electron carriers migrate outward, to line the outer part of the membrane. For bacteria and archaea, this means lining the cell membrane and explains the importance for the negative charge of the cell.

As the positively charged protons accumulate, a concentration gradient of protons develops. This results in the cytoplasm of the cell being more alkaline and more negative, leading to both a chemical and electrical potential difference. This proton motive force (PMF) can be used to do work for the cell, such as in the rotation of the bacterial flagellum or the uptake of nutrients.

ATP SYNTHASE

The PMF can also be used to synthesize ATP, with the help of an enzyme known as ATP synthase (or ATPase). This large enzyme has two components, one that spans the membrane and one that sticks into the cytoplasm and synthesizes the ATP. Protons are driven through the membrane-spanning component, generating torque that drives the rotation of the cytoplasmic portion. When the cytoplasmic component returns to its original configuration it binds Pi to ADP, generating a molecule of ATP.

CONCISE SUMMARY: AEROBIC RESPIRATION

After all that, what did the cell end up with, from using aerobic respiration? Using substrate-level phosphorylation the cell generated 2 net molecules of ATP during glycolysis, in addition to 2 molecules of ATP-equivalents from the TCA cycle. For reduced carriers, there were 2 molecules of NADH generated during glycolysis, in addition to 8 molecules from the TCA cycle or its connecting reaction. There were also 2 molecules of $FADH_2$ from the TCA cycle. All of those electrons were passed on to the ETC (and eventually to oxygen), in order to develop a PMF, so that ATP synthase could generate ATP. How much ATP is generated?

Research indicates that the process is not completely efficient and there is some "leakage" that occurs. Current estimates are that 2.5 ATP are generated for every molecule of NADH, while 1.5 ATP are generated for every molecule of $FADH_2$. Using these values would allow the cell to synthesize 25 molecules of ATP from all the NAD+ that was reduced in the process, in addition to 3 molecules of ATP from the FAD+ that was reduced. This would bring the grand total of maximum ATP produced to 32 (counting the GTP in that figure).

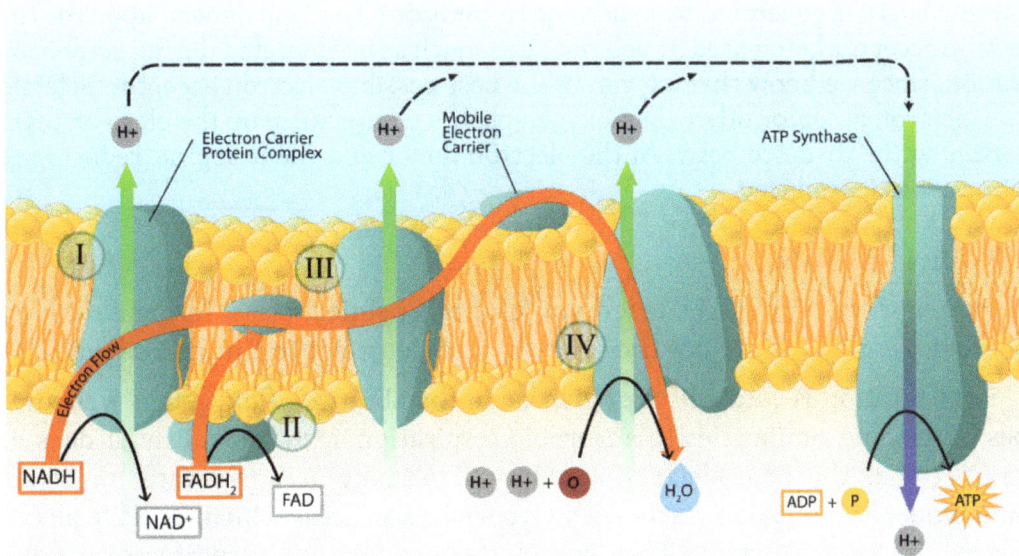

ATP Generation.

ANAEROBIC CHEMOORGANOTROPHY

Certainly oxygen is a wonderful final electron acceptor, particularly when paired with glucose as an initial electron donor. It is part of the lowest redox couple on an electron tower, with an extremely positive standard electron potential. But what does a microbe do, if oxygen is not available or it lacks the protections necessary from toxic oxygen by-products? Let us focus on the generation of energy in the absence of oxygen, using a different electron acceptor, when an organic chemical is still being used as the initial electron donor. Examples of anaerobic chemoorganotrophy include anaerobic respiration and fermentation.

ANAEROBIC RESPIRATION

Anaerobic respiration starts with glycolysis as well and the pyruvate can be shunted off to the TCA cycle, just like in aerobic respiration. In fact, oxidative phosphorylation is used to generate most of the ATP, which means the use of an ETC and ATP synthase. The key difference is that the final electron acceptor will not be oxygen.

There are a variety of possible final electron acceptors that can be used in anaerobic respiration, allowing microbes to live in a wide variety of locations. The best electron acceptor will be the one that is lowest down on the electron tower, in an oxidized form (i.e. on the left-hand side of the redox couple). Some common electron acceptors include nitrate (NO_3-), ferric iron (Fe_3+), sulfate (SO_42-), carbonate (CO_32-) or even certain organic compounds, like fumarate.

How much ATP is generated by anaerobic respiration? That will depend upon the final electron acceptor being used. It will not be as much as is generated during aerobic respiration, since we know that oxygen in the best possible electron acceptor. Selection of an electron acceptor other than oxygen pushes an organism up the electron tower, shortening the distance between the electron donor and the acceptor, reducing the amount of ATP produced.

FERMENTATION

No matter what they might teach you in a biochemical class, fermentation and anaerobic respiration are not the same thing, at least not to a microbiologist.

Fermentation is catabolism of glucose in the absence of oxygen as well and it does have some similarities to anaerobic respiration. Most obviously, it does not use oxygen as the final electron acceptor. It actually uses pyruvate, an organic compound. Fermentation starts with glycolysis, a process which we have already covered, that also starts off both aerobic respiration and anaerobic respiration. What does it yield? Two net molecules of ATP by substrate-level phosphorylation and 2 molecules of NADH. Organisms doing either aerobic or anaerobic respiration would then utilize oxidative phosphorylation in order to increase their ATP yield. Fermenters, however, lack an ETC or repress synthesis of their ETC when oxygen is not available, so they do not use the TCA cycle at all.

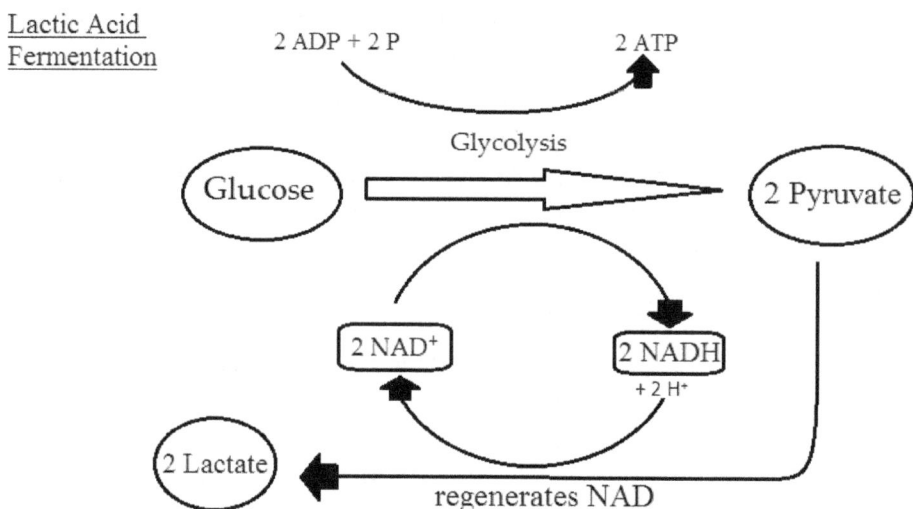

Lactate Fermentation. By Sjantoni (Own work) [CC BY-SA 3.0], via Wikimedia Commons

Without the use of an ETC (or a PMF or ATP synthase), no further ATP is generated beyond what was synthesized during glycolysis. But organisms using fermentation cannot just stop with glycolysis, since eventually all their molecules of NAD+ would become reduced. In order to re-oxidize this electron carrier they use pyruvate as a final electron acceptor, yielding a variety of fermentation products such as ethanol, CO_2, and various acids.

Fermentation products, although considered waste products for the cell, are vitally important for humans. We rely on the process of fermentation to produce a variety of fermented foods (beer, wine, bread, cheese, tofu), in addition to using the products for a variety of industrial processes.

KEY WORDS

Chemoorganotrophy, aerobic respiration, glycolysis, substrate-level phosphorylation, tricarboxylic acid (TCA) cycle, GTP, oxidative phosphorylation, proton motive force (PMF), ATP synthase/ATPase, anaerobic respiration, fermentation.

14 | Chemolithotrophy and Nitrogen Metabolism

CHEMOLITHOTROPHY

Chemolithotrophy is the oxidation of inorganic chemicals for the generation of energy. The process can use oxidative phosphorylation, just like aerobic and anaerobic respiration, but now the substance being oxidized (the electron donor) is an inorganic compound. The electrons are passed off to carriers within the electron transport chain, generating a proton motive force that is used to generate ATP with the help of ATP synthase.

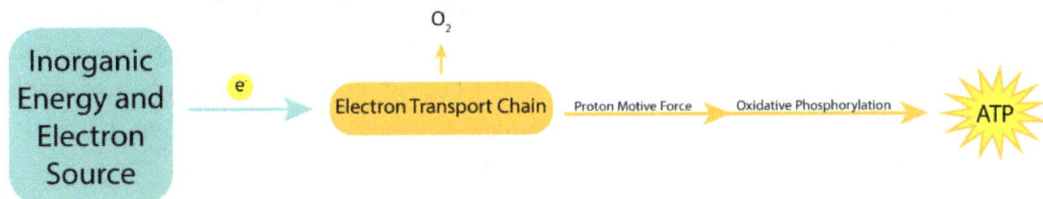

Chemolithotrophy Pathways.

Electrons donors

Chemolithotrophs use a variety of inorganic compounds as electron donors, with the most common substances being hydrogen gas, sulfur compounds (such as sulfide and sulfur), nitrogen compounds (such as ammonium and nitrite), and ferrous iron.

- **Hydrogen oxidizers** – these organisms oxidize hydrogen gas (H_2) with the use of a hydrogenase enzyme. Both aerobic and anaerobic hydrogen oxidizers exist, with the aerobic organisms eventually reducing oxygen to water.

- **Sulfur oxidizers** – as a group these organisms are capable of oxidizing a wide variety of reduced and partially reduced sulfur compounds such as hydrogen sulfide (H_2S), elemental sulfur (S_0), thiosulfate ($S_2O_3^{2-}$), and sulfite (SO_3^{2-}). Sulfate (SO_4^{2-}) is frequently a by-product of the oxidation. Often the oxidation occurs in a stepwise fashion with the help of the sulfite oxidase enzyme.

⊙ **Nitrogen oxidizers** – the oxidation of ammonia (NH3) is performed as a two-step process by nitrifying microbes, where one group oxidizes ammonia to nitrite (NO-2) and the second group oxidizes the nitrite to nitrate (NO3). The entire process is known as nitrification and is performed by small groups of aerobic bacteria and archaea, often found living together in soil or in water systems.

⊙ **Iron oxidizers** – these organisms oxidize ferrous iron (Fe+2) to ferric iron (Fe+3). Since Fe+2 has such a positive standard reduction potential, the bioenergetics are not extremely favorable, even using oxygen as a final electron acceptor. The situation is made more difficult for these organisms by the fact that Fe2+ spontaneously oxidizes to Fe3+ in the presence of oxygen; the organisms must use it for their own purposes before that happens.

ELECTRON ACCEPTORS

Chemolithotrophy can occur aerobically or anaerobically. Just as with either type of respiration, the best electron acceptor is oxygen, to create the biggest distance between the electron donor and the electron acceptor. Using a non-oxygen acceptor allows chemolithotrophs to have greater diversity and the ability to live in a wider variety of environments, although they sacrifice energy production.

AMOUNT OF ATP GENERATED

Just as both the electron donors and acceptors can vary widely for this group of organisms, the amount of ATP generated for their efforts will vary widely as well. They will not make as much ATP as an organism using aerobic respiration, since the largest $\Delta E0$' is found using glucose as an electron donor and oxygen as an electron acceptor. But how much less than 32 molecules of ATP greatly depends upon the actual donor and acceptor being used. The smaller the distance between the two, the less ATP that will be formed.

CHEMOLITHOAUTOTROPHS VS CHEMOLITHOHETEROTROPHS

Most chemolithotrophs are autotrophs (chemolithoautotrophs), where they fix atmospheric carbon dioxide to assemble the organic compounds that they need. These organisms require both ATP and reducing power (i.e. NADH/NADPH) in order to ultimately convert the oxidized molecule CO2 into a greatly reduced organic compound, like glucose. If a chemolithoautotroph is using an electron donor with a higher redox potential than NAD+/NADP, they must use reverse electron flow to push electrons back up the electron tower. This is energetically unfavorable to the cell, consuming energy from the proton motive force to drive electrons in a reverse direction back through the ETC.

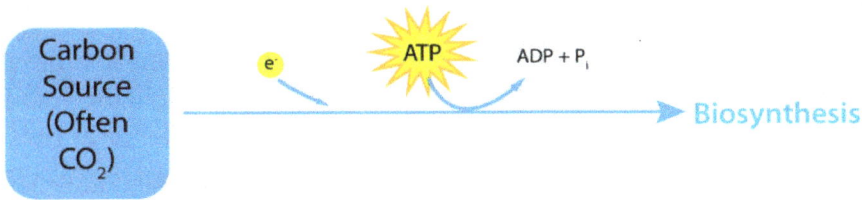

CO_2 Fixation.

Some microbes are chemolithoheterotrophs, using an inorganic chemical for their energy and electron needs, but relying on organic chemicals in the environment for their carbon needs. These organisms are also called mixotrophs, since they require both inorganic and organic chemical compounds for their growth and reproduction.

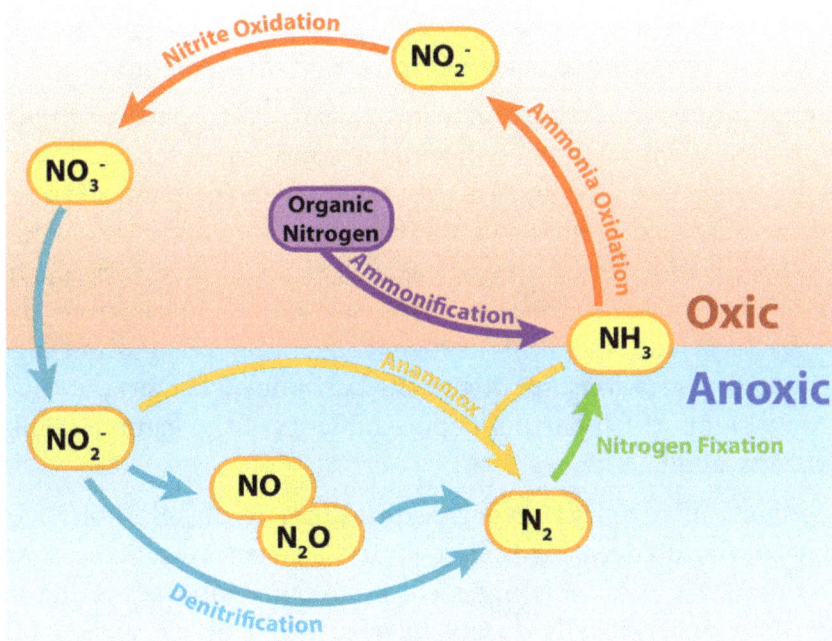

Nitrogen Cycle.

NITROGEN METABOLISM

The nitrogen cycle depicts the different ways in which nitrogen, an essential element for life, is used and converted by organisms for various purposes. Much of the chemical conversions are performed by microbes as part of their metabolism, performing a valuable service in the process for other organisms in providing them with an alternate chemical form of the element.

NITROGEN FIXATION

Nitrogen fixation describes the conversion of the relatively inert dinitrogen gas (N2) into ammonia (NH3), a much more useable form of nitrogen for most life forms. The process is performed by diazotrophs, a limited number of bacteria and archaea that can grow without an external source of fixed nitrogen, because of their abilities. Nitrogen fixation is an essential process for Earth's organisms, since nitrogen is a required component of various organic molecules, such as amino acids and nucleotides. Plants, animals, and other organisms rely on bacteria and archaea to provide nitrogen in a fixed form, since no eukaryote is known that can fix nitrogen.

Nitrogen fixation is an extremely energy and electron intensive process, in order to break the triple bond in N2 and reduce it to NH3. It requires a particular enzyme known as nitrogenase, which is inactivated by O2. Thus, nitrogen fixation must take place in an anaerobic environment. Aerobic nitrogen-fixing organisms must devise special conditions or arrangements in order to protect their enzyme. Nitrogen-fixing organisms can either exist independently or pair up with a plant host:

- ◉ Symbiotic nitrogen-fixing organisms: these bacteria partner up with a plant, to provide them with an environment appropriate for the functioning of their nitrogenase enzyme. The bacteria live in the plant's tissue, often in root nodules, fixing nitrogen and sharing the results. The plant provides both the location to fix nitrogen, as well as additional nutrients to support the energy-taxing process of nitrogen fixation. It has been shown that the bacteria and the host exchange chemical recognition signals that facilitate the relationship. One of the best known bacteria in this category is Rhizobium, which partners up with plants of the legume family (clover, soybeans, alfalfa, etc).

- ◉ Free-living nitrogen-fixing organisms: these organisms, both bacteria and archaea, fix nitrogen for their own use that ends up being shared when the organisms dies or is ingested. Free-living nitrogen-fixing organisms that grow anaerobically do not have to worry about special adaptations for their nitrogenase enzyme. Aerobic organisms must make adaptations. Cyanobacteria, a multicellular bacterium, make specialized cells known as heterocysts in which nitrogen fixation occurs. Since Cyanobacteria produce oxygen as part of their photosynthesis, an anoxygenic version occurs within the heterocyst, allowing the nitrogenase to remain active. The heterocysts share the fixed nitrogen with surrounding cells, while the surrounding cells provide additional nutrients to the heterocysts.

ASSIMILATION

Assimilation is a reductive process by which an inorganic form of nitrogen is reduced to organic nitrogen compounds such as amino acids and nucleotides, allowing for cellular growth and reproduction. Only the amount needed by the cell is reduced. Ammonia assimilation occurs when the ammonia (NH_3)/ammonium ion (NH_4^+) formed during nitrogen fixation is incorporated into cellular nitrogen. Assimilative nitrate reduction is a reduction of nitrate to cellular nitrogen, in a multi-step process where nitrate is reduced to nitrite then ammonia and finally into organic nitrogen.

NITRIFICATION

As mentioned above, nitrification is a 2-step process performed by chemolithotrophs using a reduced or partially reduced form of nitrogen as an electron donor to obtain energy. One group of chemolithotrophs can perform the first part of the nitrification process, ammonia oxidation, while a different group of chemolithotrophs can perform the nitrite oxidation that occurs in the second part of nitrification. A non-nitrogen compound would serve as the electron acceptor. ATP is gained by the process of oxidative phosphorylation, using an ETC, PMF, and ATP synthase.

DENITRIFICATION

Denitrification refers to the reduction of NO_3^- to gaseous nitrogen compounds, such as N_2. Denitrifying microbes perform anaerobic respiration, using NO_3^- as an alternate final electron acceptor to O_2. This is a type of dissimilatory nitrate reduction where the nitrate is being reduced during energy conservation, not for the purposes of making organic compounds. This produces large amounts of excess byproducts, resulting in the loss of nitrogen from the local environment to the atmosphere.

ANAMMOX

Anammox or anaerobic ammonia oxidation is performed by marine bacteria, relatively recently discovered, that utilize nitrogen compounds as both electron acceptor and electron donor as a way for the cell to generate energy. In this chemolithotrophic reaction, ammonia is oxidized anaerobically as the electron donor while nitrite is utilized as the electron acceptor, with dinitrogen gas produced as a byproduct. The reactions occur within the anammoxosome, a specialized cytoplasmic structure which constitutes 50-70% of the total cell volume. Just like denitrification, the anammox reaction removes fixed nitrogen from a local environment, releasing it to the atmosphere.

Key Words

Chemolithotrophy, hydrogen oxidizers, hydrogenase, sulfur oxidizers, sulfite oxidase, nitrogen oxidizers, nitrification, iron oxidizers, chemolithoautotroph, reverse electron flow, chemolithoheterotroph, mixotroph, nitrogen fixation, diazotroph, nitrogenase, symbiotic nitrogen-fixing organisms, Rhizobium, legume, free-living nitrogen-fixing organisms, Cyanobacteria, heterocyst, assimilation, ammonia assimilation, assimilative nitrate reduction, denitrification, dissimilatory nitrate reduction, anammox, anaerobic ammonia oxidation, anammoxosome.

15 | Introduction to Phototrophy

PHOTOAUTOTROPHS VS PHOTOHETEROTROPHS

Phototrophy (or "light eating") refers to the process by which energy from the sun is captured and converted into chemical energy, in the form of ATP. The term photosynthesis is more precisely used to describe organisms that both convert sunlight into ATP (the "light reaction") but then also proceed to use the ATP to fix carbon dioxide into organic compounds (the Calvin cycle). These organisms are the photoautotrophs. In the microbial world, there are also photoheterotrophs, organisms that convert sunlight into ATP but utilize pre-made organic compounds available in the environment. The ATP could then be used for other purposes.

PIGMENTS

In order to convert energy from sunlight into ATP, organisms use light-sensitive pigments. Plants and algae utilize chlorophylls, which are used by cyanobacteria as well. Chlorophylls are green in color, due to the fact that they absorb red and blue wavelengths ($\approx$675 nm and 430 nm) and transmit green light. The purple and green bacteria have bacteriochlorophylls, which absorb higher wavelengths ($\approx$870 nm) than the chlorophylls, allowing different phototrophs to occupy the same environment without competing with one another.

Phototrophs can contain accessory pigments as well, such as the carotenoids and phycobiliproteins. Carotenoids, which absorb blue light (400-550 nm), are typically yellow, orange, or red in color. The phycobiliproteins can be split in two groups: phycoerythrin, which transmits a red color, and phycocyanin, which transmits a blue color. The accessory pigments can serve to expand the wavelength range of light being absorbed, allowing better utilization of light available. In addition, these pigments can serve a protective function for the organism by acting as an antioxidant.

In bacteria and archaea, the phototrophic pigments are housed within invaginations of the cell membrane or within a chlorosome. Light-harvesting pigments form antennae, which funnel the light to other molecules in reaction centers, which actually perform the conversion of light energy into ATP.

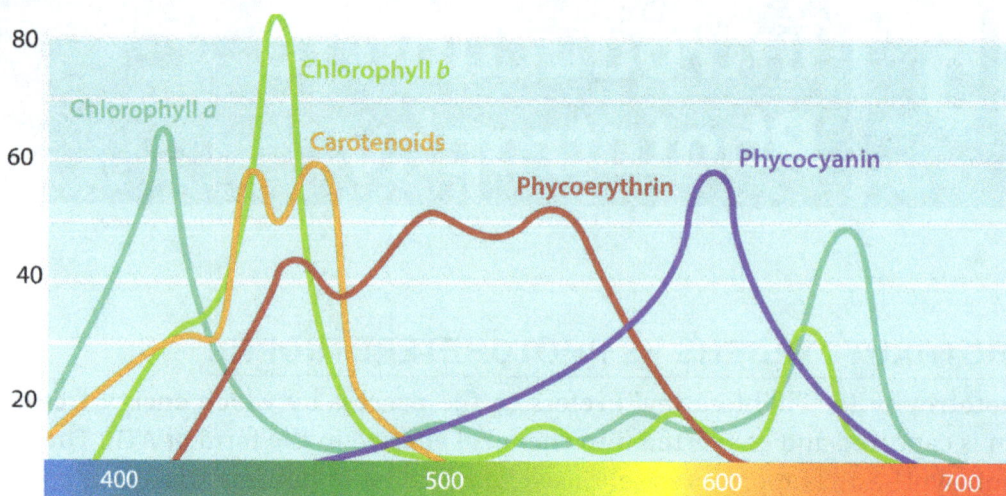

Phototrophic Pigment.

PHOTOPHOSPHORYLATION IN GENERAL

For any organism, the general process of phototrophy is going to be the same. A photosystem antennae absorbs light and funnels the energy to a reaction center, specifically to a special pair of chlorophyll/bacteriochlorophyll molecules. The molecules become excited, changing to a more negative reduction potential (i.e. jumping up the electron tower). The electrons can then be passed through an electron transport chain of carriers, such as ferredoxin and cytochromes, allowing for the development of a proton motive force.

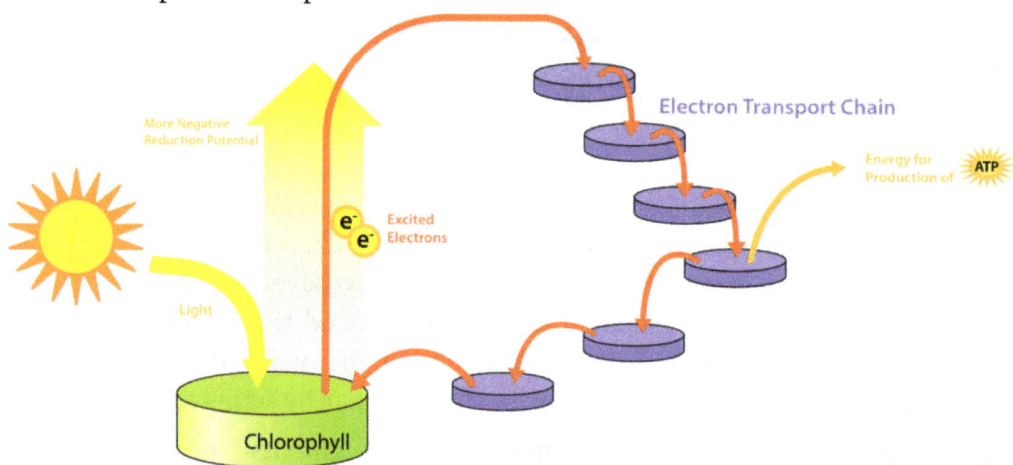

Cyclic Photophosphorylation

Anoxygenic Phototrophy

The protons are brought back across the plasma membrane through ATPase, generating ATP in the process. Since the original energy from the process came from sunlight, as opposed to a chemical, the process is called photophosphorylation. If the electrons are returned to the special pair of chlorophyll/bacteriochlorophyll molecules (cyclic photophosphorylation), the process can be repeated over and over again. If the electrons are diverted elsewhere, such as for the reduction of NAD(P) (non-cyclic photophosphorylation), then an external electron source must be used to replenish the system.

PURPLE PHOTOTROPHIC BACTERIA

Purple phototrophic bacteria engage in anoxygenic phototrophy, indicating that they do not generate oxygen during the process. They have a single photosystem with bacteriochlorophyll, allowing them to use cyclic photophosphorylation as described above for the formation of ATP. But if a purple bacterium wants to grow as a photoautotroph, it will also need reducing power in the form of NAD(P)H.

The reaction center of purple bacteria (known as P870) has an E0' of +0.5V. After being hit by a photon of light, the potential changes to -1.0V, which is insufficient to reduce NAD(P) with its E0' of -0.32V. Thus, autotrophic purple bacteria must engage in a process known as reverse electron flow, using energy from the proton motive force to drive electrons up the electron tower. Additionally, they must find an external electron donor to replenish the electrons now diverted to NAD(P). Typically the electrons come from H2S or elemental sulfur, with various sulfur byproducts produced.

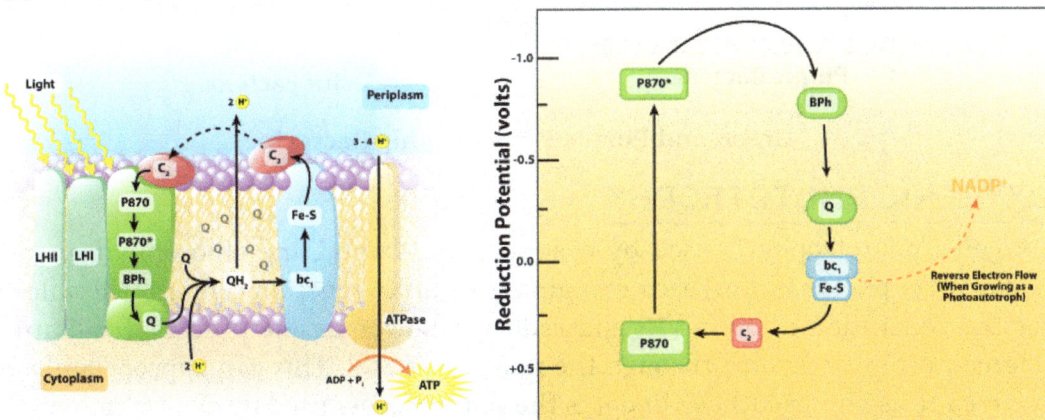

Photophosphorylation in Purple Bacteria.

In the presence of organic compounds, the purple bacteria often exist as photoheterotrophs, utilizing cyclic photophosphorylation to generate ATP and getting their organic compounds from the environment. This eliminates the need for using reverse electron flow, an energetically unfavorable process, as well as the need for external electron donors.

GREEN PHOTOTROPHIC BACTERIA

Green phototrophic bacteria also engage in anoxygenic phototrophy, utilizing a single photosystem with bacteriochlorophyll for cyclic photophosphorylation in the production of ATP. However, they also use this same photosystem for generation of reducing power, by periodically drawing off electrons to NAD+. The use of reverse electron flow is unnecessary, however, since the initial carrier, ferredoxin (Fd) has a E0' with a more negative reduction potential than NAD(P). An external electron donor is required, typically by using H2S or thiosulfate. Thus, the green bacteria operate as photoautotrophs, by alternating the use of their photosystem for ATP or NAD(P)H.

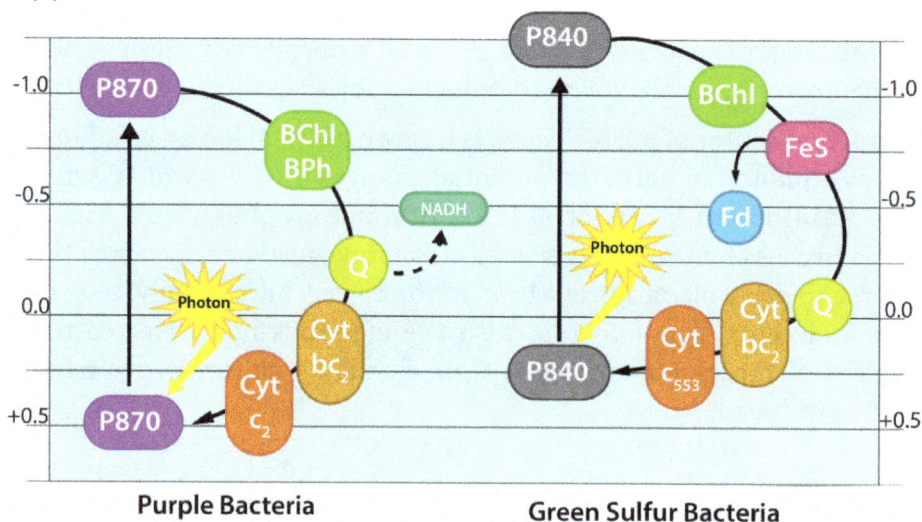

Green and Purple Phototrophic Bacteria.

OXYGENIC PHOTOTROPHY

Oxygenic phototrophy is used by cyanobacteria containing chlorophyll a, with two distinct photosystems, each containing separate reaction centers. This allows for the generation of both ATP and reducing power in one process, facilitating photoautotrophic growth through the fixation of CO2. This can appropriately be referred to as photosynthesis and it is the same process used by plants, commonly referred to as the "Z pathway."

The process starts when light energy decreases the reduction potential of P680 chlorophyll a molecules contained in photosystem II (PSII). The electrons are then passed through an electron transport chain, generating ATP via a proton motive force. Electrons are then passed to photosystem I (PSI), where they get hit by another photon of light, decreasing their reduction potential even more. The electrons are then passed through a different electron transport chain, eventually being passed off to NADP+ for the formation of NADPH.

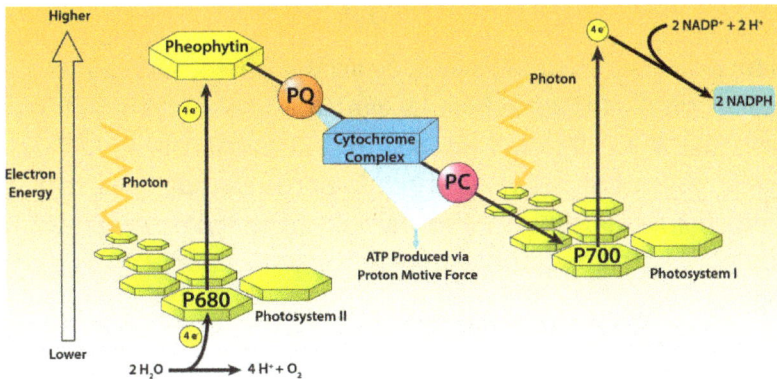

Overview of Oxygenic Photosynthesis in Cyanobacteria.

The process is an example of noncyclic photophosphorylation, since the electrons are not returned to the original photosystem. Thus, an external electron donor is required in order to allow the process to repeat. Water, found on the right side of the redox couple O2/H2O, is normally a poor electron donor, due to its extremely positive reduction potential. But the reduction potential of P680 chlorophyll a is even more positive when not excited, allowing for water to serve as an electron donor. The hydrolysis of water leads to the evolution of oxygen, a welcome byproduct for all organisms that use aerobic respiration. It is thought that cyanobacteria are responsible for the oxygenation of Earth, allowing for the development of aerobic respiration as a form of metabolism.

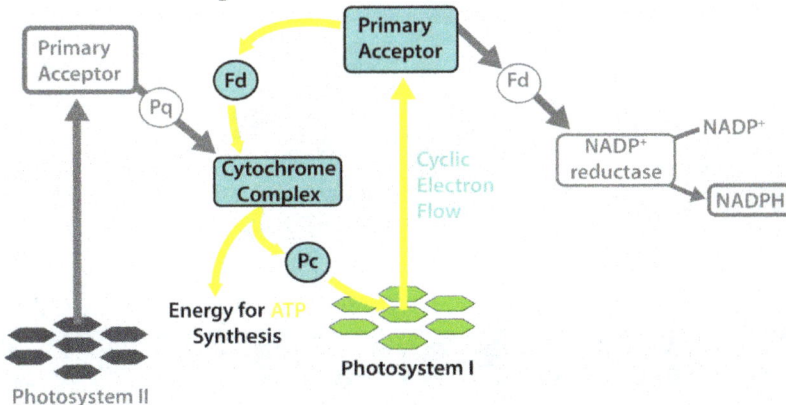

Oxygen-free Phototrophy in Cyanobacterial Heterocysts.

There are some conditions under which cyanobacteria only use PSI, essentially performing a form of anoxygenic phototrophy, despite their possession of chlorophyll a. This occurs within the heterocysts of cyanobacteria, where oxygen inactivates the nitrogenase enzymes. Heterocysts degrade PSII, ensuring that oxygen will not be produced as a byproduct, while still allowing for the production of ATP with the remaining photosystem.

RHODOPSIN-BASED PHOTOTROPHY

One unusual form of phototrophy is used by archaea, without the use of chlorophyll or bacteriochlorophyll. Instead these organisms use a bacteriorhodopsin (more appropriately called an archaearhodopsin), a retinal molecule related to the one found in vertebrate eyes. When the rhodopsin absorbs light it undergoes a conformational change, pumping a proton across the cell membrane and leading to the development of a proton motive force, without the participation of an electron transport chain.

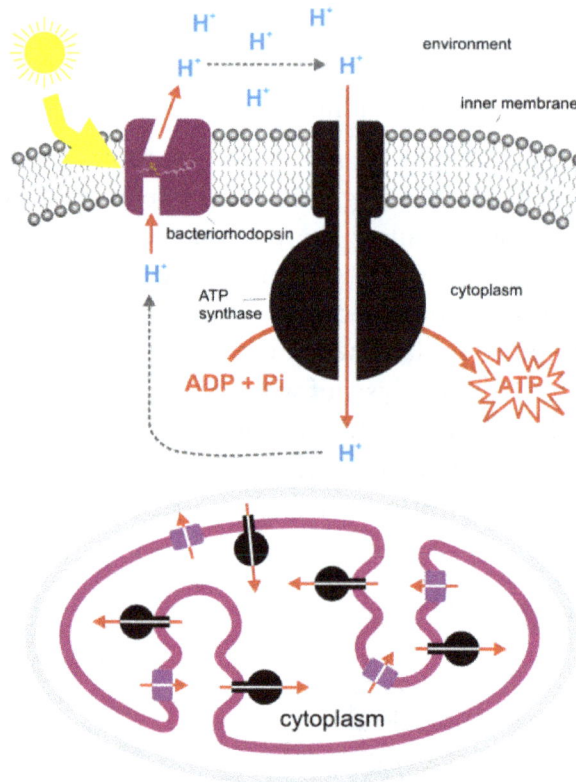

Rhodopsin-Based Phototrophy. By Darekk2 (Own work) [CC BY-SA 3.0]

KEY WORDS

Phototrophy, photosynthesis, photoautotroph, photoheterotroph, chlorophylls, bacteriochlorophylls, carotenoid, phycobiliprotein, phycoerythrin, phycocyanin, chlorosome, antennae, reaction centers, photophosphorylation, cyclic photophosphorylation, non-cyclic photophosphorylation, purple phototrophic bacteria, anoxygenic phototrophy, reverse electron flow, green phototrophic bacteria, oxygenic phototrophy, Z pathway, photosystem II (PSII), photosystem I (PSI), rhodopsin-based phototrophy, bacteriorhodopsin/archaeorhodopsin.

16 | Taxonomy and Evolution

INTRODUCTION TO EVOLUTION

It is believed that the Earth is 4.6 billion year old, with the first cells appearing approximately 3.8 billion years ago. Those cells were undoubtably microbes, eventually giving rise to all the life forms that we envision today, as well as the life forms that went extinct before we got here. How did this progression occur?

EARLY EARTH

Conditions on early Earth were most likely extremely hot, anoxic (lacking oxygen), with reduced inorganic chemicals in abundance. While no one knows exactly how cells came about, it is likely that they were initially suited to these harsh conditions.

RNA WORLD

RNA, in its many forms, plays a crucial role in cellular activities. It has been hypothesized that RNA played an even more central role in primitive cells, with self-replicating RNA containing a cell's information as well as having catalytic activity to synthesize proteins. Eventually this RNA world evolved to one in which proteins took over the catalytic responsibilities and DNA became the common form of information storage.

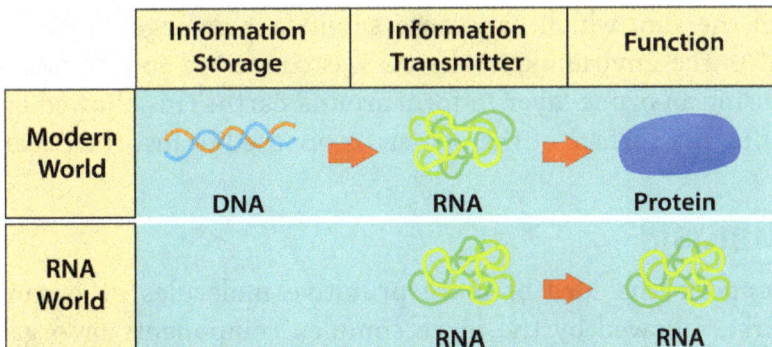

The "RNA World" and the Modern World.

METABOLIC DIVERSITY

Initial cells probably had a relatively primitive electron transference system, perhaps through just one carrier, that still allowed for the development of a proton motive force to conserve energy. As chemolithoautotrophs proliferated, organic material started to accumulate in the environment, providing the conditions needed for the development of chemoorganotrophic organisms. These new cells oxidized organic compounds, with their more negative redox potential and increased number of electrons. This most likely lengthened electron transport chains, resulting in faster growth, and speeding up diversity even more.

PHOTOTROPHY & PHOTOSYNTHESIS

At about 3.5 billion year ago some cells evolved phototrophic pigments, allowing for the conversion of light energy into chemical energy. Initially phototrophs utilized anoxygenic phototrophy, using sulfur products as an electron donor when performing CO_2 fixation.

Stromatolites are layered rocks that form when minerals are incorporated into thick mats of microbes, growing on water surfaces. Ancient stromatolites contain fossilized microbial mats made up of cyanobacteria-like cells, indicating their presence relatively early in Earth's history.

Approximately 2.5-3.3 billion year ago the cyanobacterial ancestors developed oxygenic photosynthesis by acquiring two photosystems and the pigment chlorophyll a. This led to the use of water as an electron donor, causing oxygen to accumulate in Earth's atmosphere. This Great Oxidation Event substantially changed the types of metabolism possible, allowing for the use of oxygen as a final electron acceptor.

OZONE SHIELD FORMATION

The development of an ozone shield around the Earth occurred around 2 billion years ago. Ozone (O_3) serves to block out much of the ultraviolet (UV) radiation coming from the sun, which can cause significant damage to DNA. As oxygen accumulated in the environment, the O_2 was converted to O_3 when exposed to UV light, causing an ozone layer to form around Earth. This allowed organisms to start inhabiting the surface of the planet, as opposed to just the ocean depths or soil layers.

ENDOSYMBIOSIS

Evolution supports the idea of more primitive molecules or organisms being generated first, followed by the more complex components or organisms over time. Endosymbiosis offers an explanation for the development of eukaryotic cells, a more complex cell type with organelles or membrane-bound enclosures.

It is generally accepted that eukaryotic ancestors arose when a cell ingested another cell, a free-living bacterium, but did not digest it. This endosymbiont had capabilities that the proto-eukaryotic cell lacked, such as the ability for phototrophy (i.e. chloroplasts) or oxidative phosphorylation (i.e. mitochondria). Eventually the two became mutually dependent upon one another with the endosymbiont becoming an organelle, with the chloroplast being derived from a cyanobacterial ancestor and the mitochondrion being derived from a gram negative bacillus ancestor.

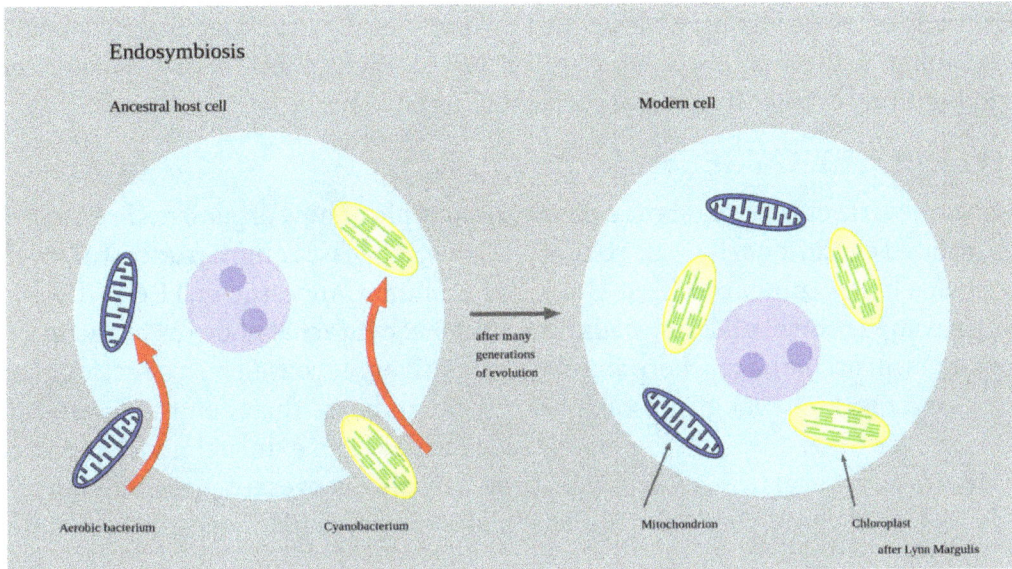

Endosymbiosis. By Signbrowser (Own work) [CC0], via Wikimedia Commons

Evidence to support this idea includes the fact that mitochondria and chloroplasts: have a single, circular chromosome; undergo binary fission separate from the eukaryotic cell; have 70S sized ribosomes; have a lipid bilayer with a 2:1 ratio of protein to lipid; and, perhaps most importantly, have rRNA sequences that place them phylogenetically with the bacteria.

PHYLOGENY

Molecular Phylogeny

Phylogeny is a reference to the development of an organism evolutionarily. Molecular techniques allow for the evolutionary assessment of organisms using genomes or ribosomal RNA (rRNA) nucleotide sequences, generally believed to provide the most accurate information about the relatedness of microbes.

Nucleic acid hybridization or DNA-DNA hybridization is a commonly used tool for molecular phylogeny, comparing the similarities between genomes. The genomes of two organisms are heated up or "melted" to separate the complementary strand and then allowed to cool down. Strands that have complementary base sequences will re-anneal, while strands without complementation will remain upaired. Typically one source of DNA is labeled, usually with radioactivity, to allow for identification of each DNA source.

Nucleic acid sequencing, typically using the rRNAs from small ribosomal subunits, allows for direct comparison of sequences. The ribosomal sequence is seen as ideal because the genes encoding it do not change very much over time, nor does it appear to be strongly influenced by horizontal gene transfer. This makes it an excellent "molecular chronometer," or way to track genetic changes over a long period of time, even between closely related organisms.

PHYLOGENETIC TREES

Phylogenetic trees serve to show a pictorial example of how organisms are believed to be related evolutionarily. The root of the tree is the last common ancestor for the organisms being compared (Last Universal Common Ancestor or LUCA, if we are doing a comparison of all living cells on Earth). Each internal node (or branchpoint) represents an occurrence where the organisms diverged, based on a genetic change in one organism. The length of each branch can indicate the amount of molecular changes over time, if the phylogenetic tree is scaled. The external nodes represent specific taxa or organisms (although they can also represent specific genes). A clade indicates a group of organisms that all have a particular ancestor in common.

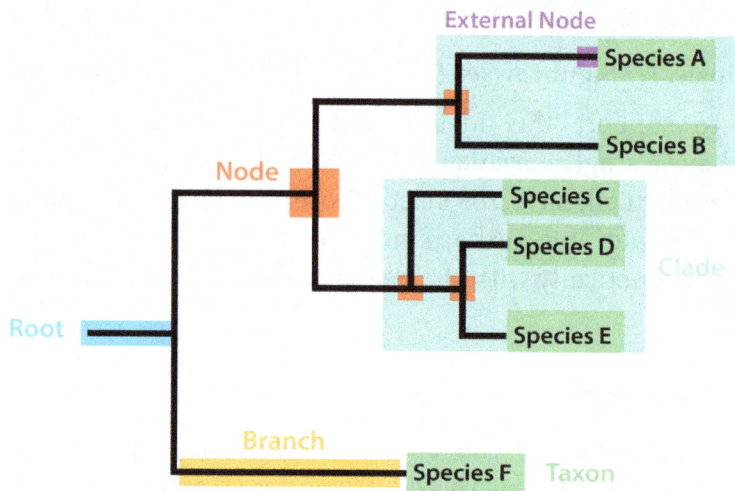

Phylogenetic Tree Structure

TAXONOMY

Taxonomy refers to the organization of organisms, based on their relatedness. Typically it involves some type of classification scheme, the identification of isolates, and the naming or nomenclature of included organisms. Many different classification schemes exist, although many have not been appropriate for comparison of microorganisms.

CLASSIFICATION SYSTEMS

A phenetic classification system relies upon the phenotypes or physical appearances of organisms. Phylogenetic classication uses evolutionary relationships of organisms. A genotypic classification compares genes or genomes between organisms. The most popular approach is to use a polyphasic approach, which combines aspects of all three previous systems.

MICROBIAL SPECIES

Currently there is no widely accepted "species definition" for microbes. The definition most commonly used is one that relies upon both genetic and phenotypic information (a polyphasic approach), with a threshold of 70% DNA-DNA hybridization and 97% 16S rRNA sequence identity in order for two organisms to be deemed as belonging to the same species.

KEY WORDS
Evolution, RNA world, stromatolites, Great Oxidation Event, ozone shield, endosymbiosis, chloroplast, mitochondria, phylogeny, ribosomal RNA/rRNA, molecular phylogeny, nucleic acid hybridization, DNA-DNA hybridization, nucleic acid sequencing, molecular chronometer, phylogenetic tree, Last Universal Common Ancestor/LUCA, node, branch, external node, clade, taxonomy, phonetic classification, phylogenetic classification, genotypic classification, polyphasic classiciation, species definition.

17 | Introduction to Microbial Genetics

Let's talk about sex. Bacterial sex. Ha! That is going to be difficult, since bacteria do not have sex. Which presents a real problem for bacteria (and archaea, too) – how do they get the genetic variability that they need? They might need a new gene to break down an unusual nutrient source or degrade an antibiotic threatening to destroy them – acquiring the gene could mean the difference between life and death. But where would these genes come from? How would the bacteria get a hold of them? We are going to explore the processes that bacteria use to acquire new genes, the mechanisms known as Horizontal Gene Transfer (HGT).

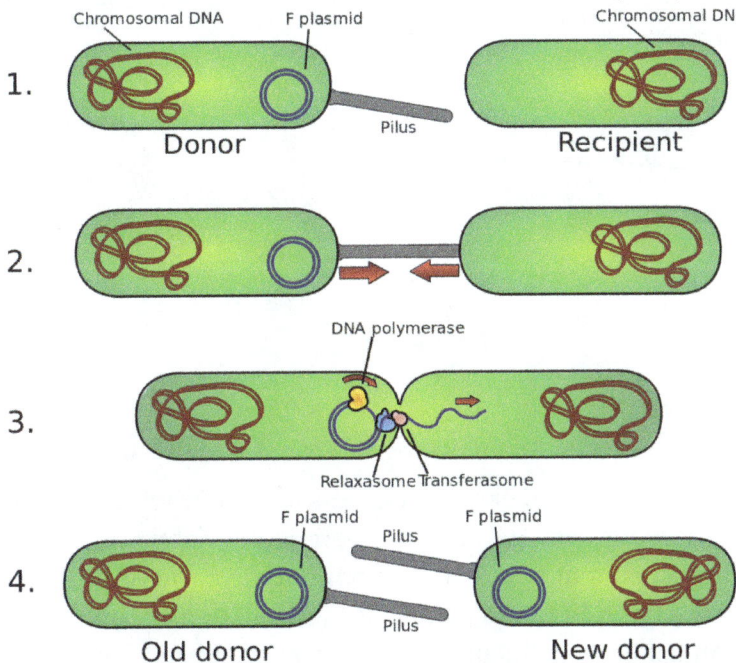

Conjugation. By Adenosine (Own work) [CC BY-SA 3.0], via Wikimedia Commons

CONJUGATION

Conjugation is the process by which a donor bacterium transfers a copy of a plasmid to a recipient bacterium, through a pilus. The process requires cell-to-cell contact. The donor cell (F+) has a conjugative plasmid, an extrachromosomal piece of dsDNA that codes for the proteins necessary to make a threadlike filament known as a pilus. The pilus is used to bind to the recipient (F-) cell, bringing it in close proximity to the donor cell. It is believed that a channel is then opened between the two cells, allowing for a ssDNA copy of the plasmid to enter the recipient cells. Both cells then make the complementary copy to the ssDNA, resulting in two F+ cells capable of conjugation.

TRANSFORMATION

The process of transformation also allows a bacterial cell to acquire new genes, but it does not require cell-to-cell contact. In this process the new genes are acquired directly from the environment. Typically the process requires a donor cell that at some point lysed and released naked DNA to the environment. The recipient cell is one that is capable of taking up the DNA from the environment and incorporating it into its own genome, where the cell is described as being competent.

Transformation Summary

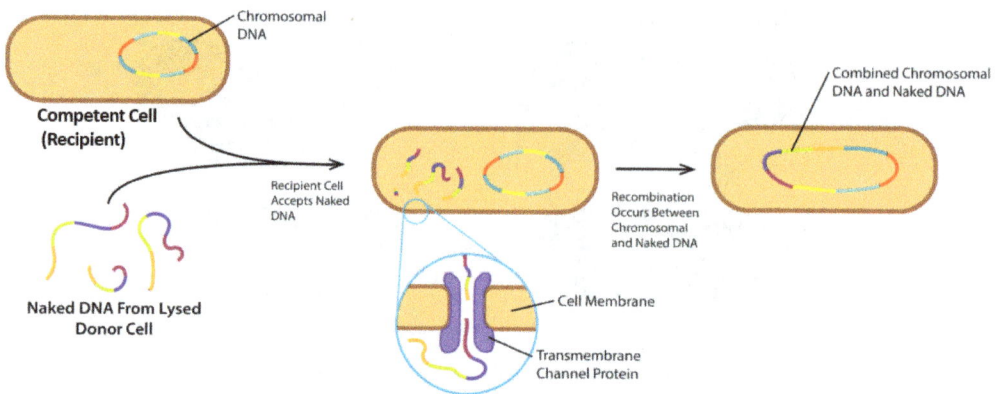

Gene Acquisition via Transformation.

There are mechanical and chemical means of encouraging a cell to pick up DNA from the environment, but natural competence is determined genetically. The process typically occurs at the end of exponential phase of growth or beginning of the stationary phase, in the presence of high cell density and limited nutrients. Under these conditions specific proteins are manufactured including DNA-binding proteins (DNA translocase), endonucleases, and transmembrane channel proteins. Gram negative cells also make a cell wall autolysin, to transport the DNA across the outer membrane.

Random pieces of DNA bind to receptors on the outside of the cell and are then transported into the cell by the DNA translocase, through the transmembrane channel, a large structure often involving numerous different proteins. An endonuclease can be used to degrade one strand of dsDNA, if only ssDNA may pass into the cell, or to cleave the DNA fragment into smaller sizes .Once inside the cell, the DNA must be incorporated into the bacterial chromosome by RecA (see Molecular Recombination below), for the genes to be expressed.

TRANSDUCTION

Transduction involves the use of a virus, a bacteriophage, to act as a conduit for shuttling bacteria genes from one cell to another, thus negating the necessity for cell-to-cell contact. There are two different types of transduction: generalized transduction and specialized transduction.

GENERALIZED TRANSDUCTION

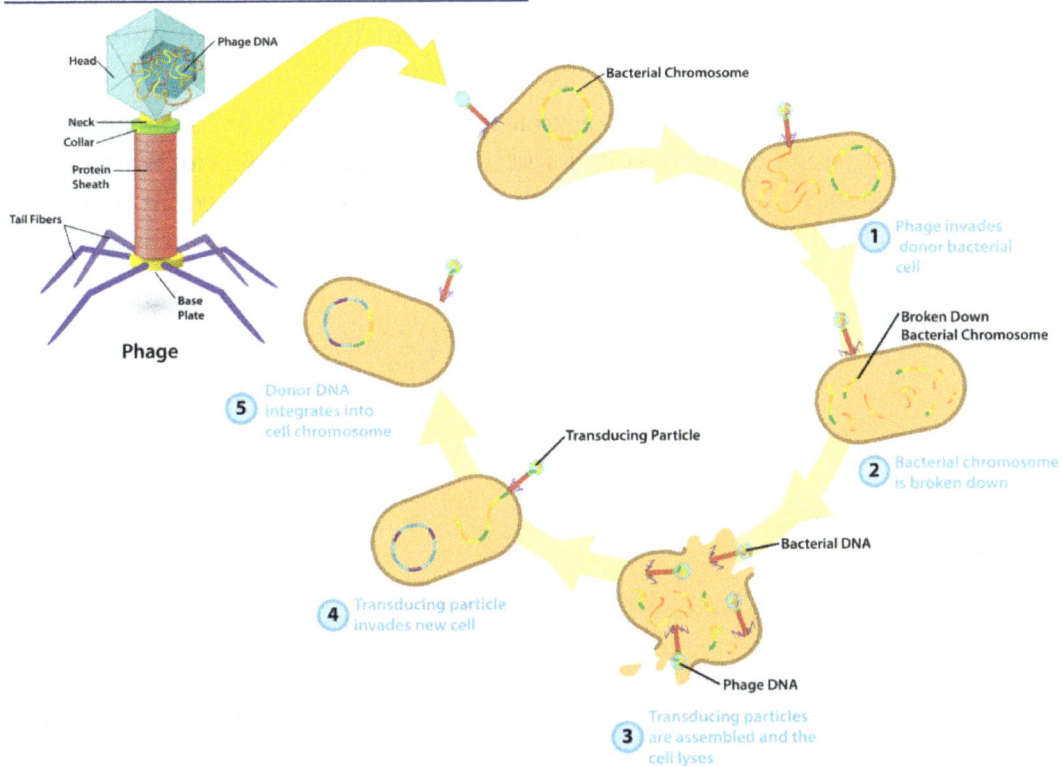

Generalized Transduction.

In generalized transduction, a bacterial host cell is infected with either a virulent or a temperate bacteriophage engaging in the lytic cycle of replication. After the first three steps of replication (absorption, penetration, and synthesis), the virus enters into the assembly stage, during which fully formed virions are made. During this stage, random pieces of bacterial DNA are mistakenly packaged into a phage head, resulting in the production of a transducing particle. While these particles are not capable of infecting a cell in the conventional sense, they can bind to a new bacterial host cell and inject their DNA inside. If the DNA (from the first bacterial host cell) is incorporated into the recipient's chromosome, the genes can be expressed.

SPECIALIZED TRANSDUCTION

Specialized transduction can only occur with temperate bacteriophage, since it involves the lysogenic cycle of replication. The bacteriophage randomly attaches to a bacterial host cell, injecting viral DNA inside. The DNA integrates into the chromosome of the host cell, forming a prophage. At some point induction occurs, where the prophage is excised from the bacterial chrosomsome. In specialized transduction, the excision is incorrectly performed and a portion of bacterial genes immediately adjacent to the viral genes are excised too. Since this DNA is used as the template for the synthesis stage, all copies will be a hybrid of viral and bacterial DNA, and all resulting virions will contain both viral and bacterial DNA.

Once the cell is lysed, the virions are released to infect other bacterial host cells. Each virion will attach to the host cell and inject in the DNA hybrid, which could be incorporated into the host chromosome, if a prophage is formed. At this point the second bacterial host cell can contain its own DNA, DNA from the previous bacterial host cell, and viral DNA.

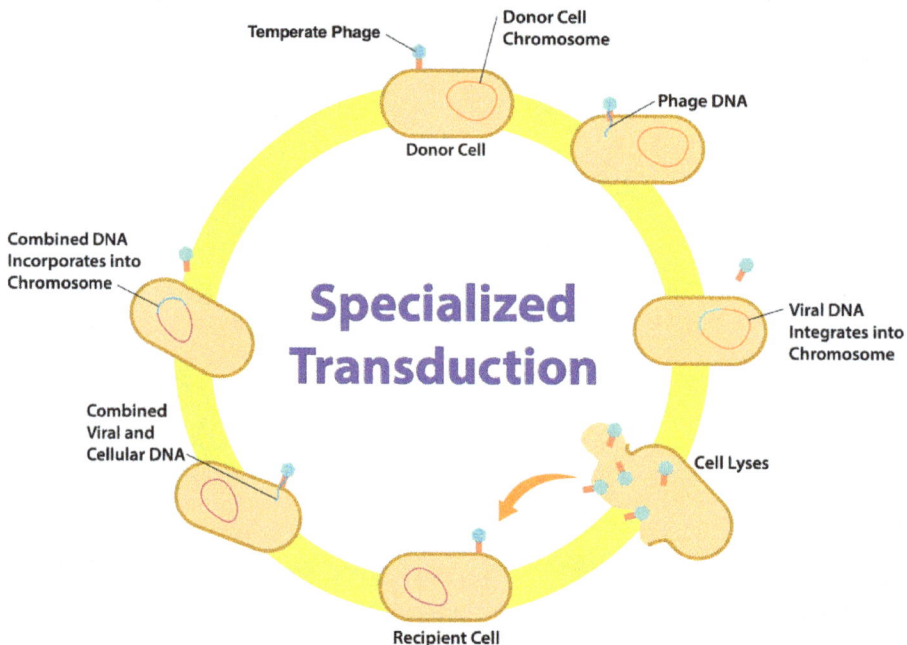

Molecular Recombination

In each of the cases of HGT, the process is only successful if the genes can be expressed by the altered cell. In conjugation, the genes are located on a plasmid, under the control of promoters on the plasmid. In transformation and transduction, where naked DNA is gaining access to the cell, the DNA could easily be broken down by the cell with no genetic expression occurring. In order for the genes to be expressed, the DNA must be recombined with the recipient's chromosome.

The most common mechanism of molecular recombination is homologous recombination, involving the RecA protein. In this process DNA from two sources are paired, based on similar nucleotide sequence in one area. An endonuclease nicks one strand, allowing RecA to pair up bases from different strands, a process known as strand invasion. The cross-over between DNA molecules is resolved with resolvase, which cuts and rejoins the DNA into two separate dsDNA molecules.

Recombination can also occur using site-specific recombination, a process often used by viruses to insert their genome into the chromosome of their host. This type of recombination is also used by transposable elements .

TRANSPOSABLE ELEMENTS

Finally, we shouldn't leave the topic of microbial genetics without at least exploring the role of transposable elements or "jumping genes." While these can play a very big role in the activation and inactivation of bacterial genes, the best explanation derives from the work of Barbara McClintock in corn, who won the Nobel Prize for her research in 1983. She demonstrated that transposable elements can be responsible for the activation or inactivation of genes within an organism.

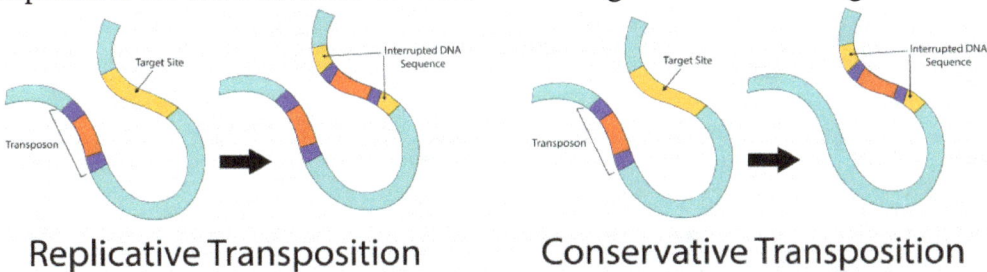

Replicative Transposition   Conservative Transposition

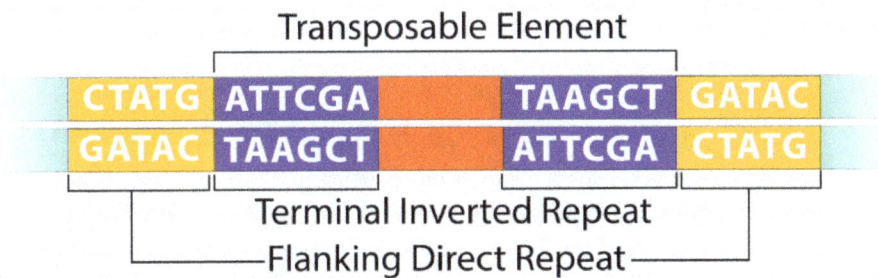

Transposition.

Transposable elements are relatively simple in structure, designed to move from one location to another within a DNA molecule by a process known as transposition. All transposable elements code for the enzyme transposase, the enzyme responsible allowing transposition to occur, and have short inverted repeats (IRs) at each end.

The simplest transposable element is an insertion sequence (IS), which contains the transposase and IRs of varying lengths. A transposon typically contains additional genes, with the exact type varying widely from transposon to transposon. A transposon can be removed from one location and relocated to another (the cut-and-paste model), a process known as conservative transposition. Alternatively, it can be copied, with the copy being inserted at a second site, in a process known as replicative transposition.

Key Words

Horizontal Gene Transfer (HGT), conjugation, donor, recipient, conjugative plasmid, F-, F, transformation, naked DNA, competence, competent cell, DNA translocase, endonuclease, autolysin, RecA, transduction, generalized transduction, transducing particle, specialized transduction, molecular recombination, homologous recombination, resolvase, site-specific recombination, transposable elements, transposition, transposase, inverted repeats (IR), insertion sequence (IS), transposon, conservative transposition, replicative transposition.

18 | Introduction to Genetic Engineering

Genetic engineering is the deliberate manipulation of DNA, using techniques in the laboratory to alter genes in organisms. Even if the organisms being altered are not microbes, the substances and techniques used are often taken from microbes and adapted for use in more complex organisms.

STEPS IN CLONING A GENE

Let us walk through the basic steps for cloning a gene, a process by which a gene of interest can be replicated many times over. Let us pretend that we are going to genetically engineer E. coli cells to glow in the dark, a characteristic that they do not naturally possess.

- Isolate DNA of interest – first we need to identify the genes or genes that we are interested in, the target DNA. If we want our E. coli cells to glow in the dark, we need to find an organism that possesses this trait and identify the gene or genes responsible for the trait. The green fluorescent protein (GFP) commonly used as an expression marker in molecular techniques was originally isolated from jellyfish.In cloning a gene it is helpful to use a cloning vector, typically a plasmid or virus, capable of independent replication that will stably carry the target DNA from one location to another. Plasmid vectors are available from both bacteria and yeast.

- Cut DNA with restriction endonucleases – once the target and vector DNA have been identified, both types of DNA are cut using restriction endonucleases. These enzymes recognize short sequences of DNA that are 4-8 bp long. The enzymes are widespread in both bacteria and archaea, with each enzyme recognizing a specific inverted repeat sequence that is palindromic (reads the same on each DNA strand, in the 5' to 3' direction).

- While some restriction endonucleases cut straight across the DNA (i.e. blunt cut), many make staggered cuts, producing a very short region of single-stranded DNA on each strand. These single-stranded regions are referred to as "sticky ends," and are invaluable in molecular cloning since the unpaired bases will recombine with any DNA having the complementary base sequence.

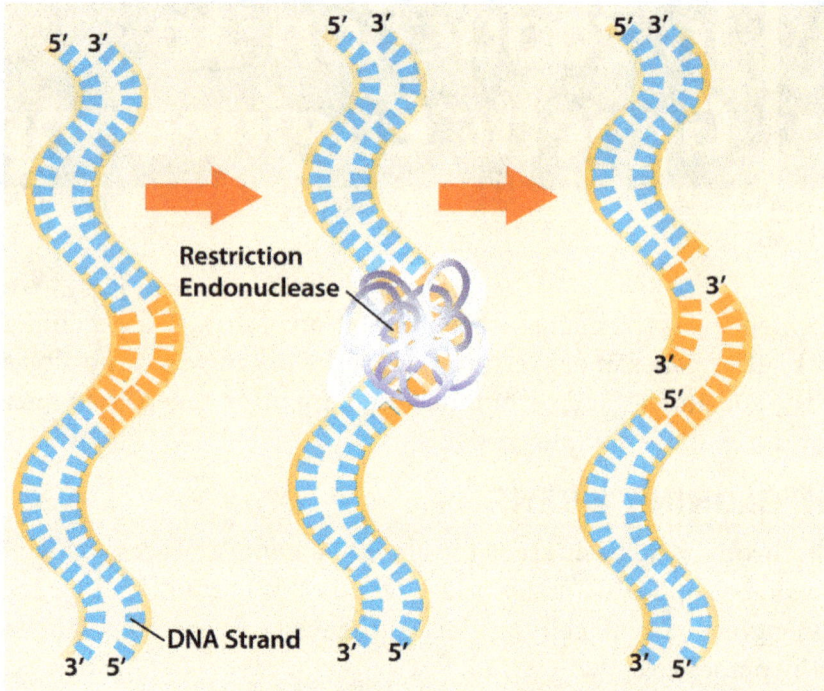

Restriction Endonucleases.

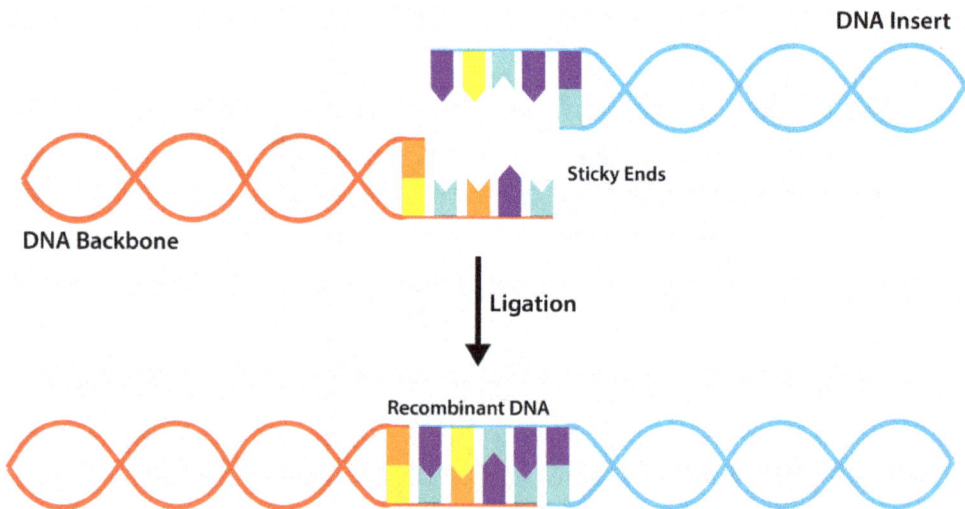

Ligation.

⊙ Combine target and vector DNA – after both types of DNA have been cleaved by the same restriction endonuclease, the two types of DNA are combined together with the addition of DNA ligase, an enzyme that repairs the covalent bonds on the sugar-phosphate backbone of the DNA.

This results in the creation of recombinant DNA, DNA molecules that contain the DNA from two or more sources, also known as chimeras. Introduce recombined molecule into host cell – once the target DNA has been stably combined with vector DNA, the recombinant DNA must be introduced into a host cell, in order for the genes to be replicated or expressed. There are different methods for introducing the recombinant DNA, largely depending upon the complexity of the host organism. In the case of bacteria, transformation is often the easiest method, using competent cells to pick up the recombinant DNA molecules. Alternatively, electroporation can be used, where the cells are exposed to a brief pulse of high –voltage electricity causing the plasma membrane to become temporarily permeable to DNA passage.While some cells will acquire recombinant DNA with the appropriate configuration (i.e. target DNA combined with vector DNA), the method also will yield cells carrying recombinant DNA with alternate DNA combinations (i.e. plasmid DNA combining with another plasmid DNA molecule or target DNA attached to more target DNA). The mixture is referred to as a genomic library and must be screened to select the appropriate clone. If random fragments of DNA were originally used (instead of isolation of the appropriate target DNA genes), the process is referred to as shotgun cloning and can yield thousands or tens of thousands of clones to be screened.

INTRODUCING RECOMBINANT DNA INTO CELLS OTHER THAN BACTERIA

Agrobacterium tumefaciens and the Ti plasmid

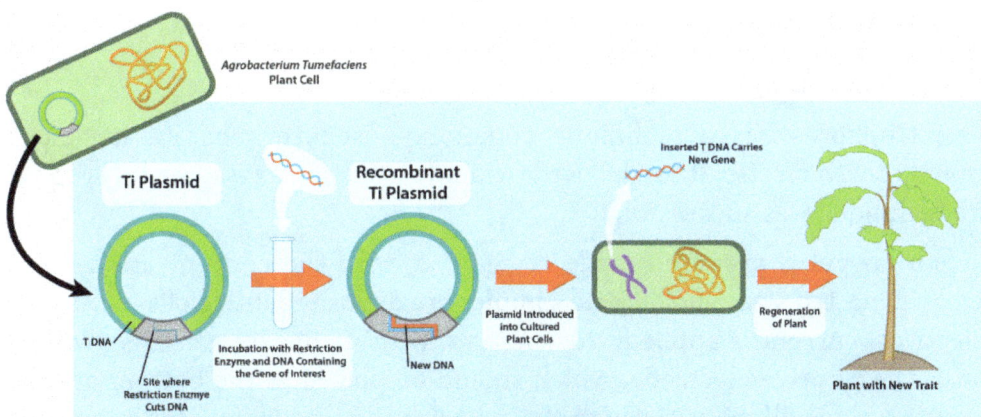

Tumor-Inducing Plasmid.

Agrobacterium tumefaciens is a plant pathogen that causes tumor formation called crown gall disease. The bacterium contains a plasmid known as the Ti (tumor inducing) plasmid, which inserts bacterial DNA into the host plant genome. Scientists utilize this natural process to do genetic engineering of plants by inserting foreign DNA into the Ti plasmid and removing the genes necessary for disease, allowing for the production of transgenic plants.

GENE GUN

A gene gun uses very small metal particles (microprojectiles) coated with the recombinant DNA, which are blasted at plant or animal tissue at a high velocity. If the DNA is transformed or taken up by the cell's DNA, the genes are expressed.

VIRAL VECTORS

For a viral vector, virulence genes from a virus can be removed and foreign DNA inserted, allowing the virus capsid to be used as a mechanism for shuttling genetic material into a plant or animal cell. Marker genes are typically added that allow for identification of the cells that took up the genes.

CRISPR-CAS9 GENOME EDITING

CRISPR stands for Clustered Regularly Interspaced Short Palindromic Repeats (now you know why everyone just says CRISPR!). It refers to a family of DNA sequences that are commonly found in bacterial and archaeal genomes, where the sequences originally came from lysogenic bacteriophage. The cells use these DNA fragments, combined with the Cas9 (CRISPR-associated protein 9) enzyme, to identify and destroy similar DNA, thus preventing subsequent bacteriophage infections. But this natural gene editing process can be used to directly edit genomes in eukaryotic organisms as well, typically by either deleting a gene or by inserting a gene.

DNA TECHNIQUES

Gel Electrophoresis

Gel electrophoresis is a technique commonly used to separate nucleic acid fragments based on size. It can be used to identify particular fragments or to verify that a technique was successful.

A porous gel is prepared made of agarose, with the concentration adjusted based on expected size. Nucleic acid samples are deposited into wells in the gel and an electrical current is applied. Nucleic acid, with its negative charge, will move towards the positive electrode, which should be placed at the bottom of the gel. The nucleic acid will move through the gel, with the smallest pieces encountering the least resistance and thus moving through the fastest. The length of passage of each nucleic acid fragment can be compared to a DNA ladder, with fragments of known size.

Polymerase Chain Reaction (PCR)

The polymerase chain reaction or PCR is a method used to copy or amplify DNA in vitro. The process can yield a billionfold copies of a single gene within a short period of time. The template DNA is mixed with all the ingredients necessary to make DNA copies: primers (small oligonucleotides that flank the gene or genes of interest by recognizing sequences on either side of it), nucleotides (the building blocks of DNA), and DNA polymerase. The steps involve heating the template DNA in order to denature or separate the strands, dropping the temperature to allow the primers to anneal, and then heating the mixture up to allow the DNA polymerase to extend the primers, using the original DNA as an initial template. The cycle is repeated 30-20 times, exponentially increasing the amount of target DNA in a few hours.

Polymerase chain reaction - PCR

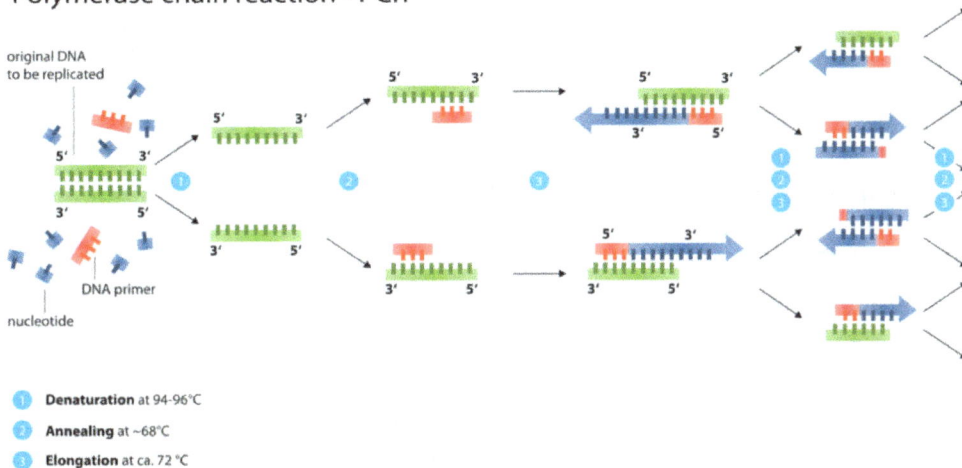

1 **Denaturation** at 94-96°C

2 **Annealing** at ~68°C

3 **Elongation** at ca. 72 °C

Polymerase Chain Reaction (PCR). By Enzoklop (Own work) [CC BY-SA 3.0], via Wikimedia Commons

Uses of Genetically Engineered Organisms

There can be numerous reasons to create a genetically modified organism (GMO) or transgenic organism, defined as a genetically modified organism that contains a gene from a different organism. Typically the hope is that the GMO will provide needed information or a product of value to society.

SOURCE OF DNA

Genetically engineered organisms can be made so that a piece of DNA can be easily replicated, providing a large source of that DNA. For example, a gene associated with breast cancer can be spliced into the genome of E. coli, allowing for the rapid production of the gene so that it may be sequenced, studied, and manipulated, without requiring repeated tissue donations from human volunteers.

SOURCE OF RNA

Antisense RNA is ssRNA that is complementary to the mRNA that will code for a protein. In cells it is made as a way to control target genes. There has been increasing interest in the use of antisense RNA as a way to prevent diseases that are caused by the production of a particular protein.

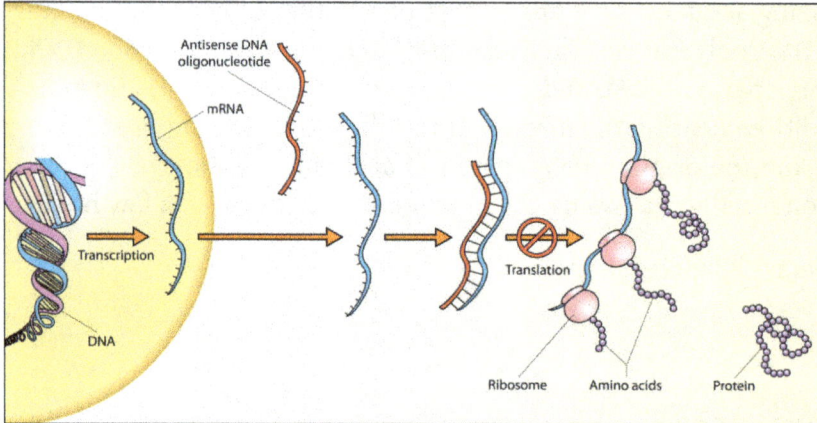

Antisense RNA. By Robinson R [CC BY 2.5], via Wikimedia Commons

Source of Protein

Since microbes replicate so rapidly, it can be extremely advantageous to use them to manufacture proteins of interest or value. Given the right promoters, bacteria will express genes for proteins that are not naturally found in bacteria, such as cytokine. Genetically engineered cells have been used to make a wide riety of proteins of use to humans, such as insulin or human growth hormone.

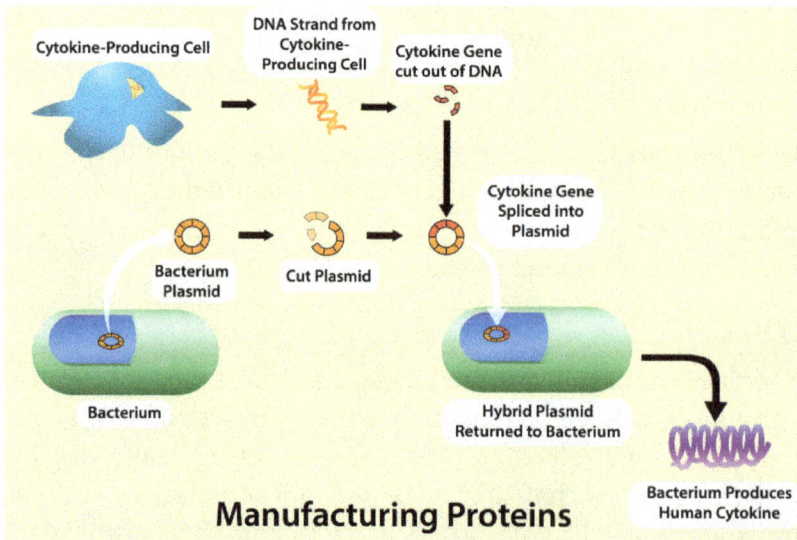

KEY WORDS

Genetic engineering, cloning, target DNA, green fluorescent protein (GFP), cloning vector, restriction endonuclease, sticky ends, DNA ligase, recombinant DNA, chimera, transformation, electroporation, genomic library, shotgun cloning, Agrobacterium tumefaciens, Ti plasmid, gene gun, microprojectiles, viral vector, CRISPR-Cas9 genome editing, gel electrophoresis, DNA ladder, polymerase chain reaction (PCR), template DNA, primer, nucleotide, DNA polymerase, denaturing, annealing, extending, genetically modified organism (GMO), transgenic organisms, antisense RNA.

19 | Introduction to Genomics

Genomics is a field that studies the entire collection of an organism's DNA or genome. It involves sequencing, analyzing, and comparing the information contained within genomes. Since sequencing has become much less expensive and more efficient, vast amounts of genomic information is now available about a wide variety of organisms, but particularly microbes, with their smaller genome size. In fact, the biggest bottleneck currently is not the lack of information but the lack of computing power to process the information!

SEQUENCING

Sequencing, or determining the base order of an organism's DNA or RNA, is often one of the first steps to finding out detailed information about an organism. A bacterial genome can range from 130 kilobase pairs (kbp) to over 14 Megabase pairs (Mbp), while a viral genome ranges from 0.859 to 2473 kbp. For comparison, the human genome contains about 3 billion base pairs.

SHOTGUN SEQUENCING

Shotgun sequencing initially involves construction of a genomic library, where the genome is broken into randomly sized fragments that are inserted into vectors to produce a library of clones. The fragments are sequenced and then analyzed by a computer, which searches for overlapping regions to form a longer stretch of sequence. Eventually all the sequences are aligned to give the complete genome sequence. Errors are reduced because many of the clones contain identical or near identical sequences, resulting in good "coverage" of the genome.

SECOND GENERATION DNA SEQUENCING

Second-generation DNA sequencing uses massively parallel methods, where multiple samples are sequenced side-by-side. DNA fragments of a few hundred bases each are amplified by PCR and then attached to small bead, so that each bead carries several copies of the same section of DNA. The beads are put into a plate containing more than a million wells, each with one bead, and the DNA fragments are sequenced.

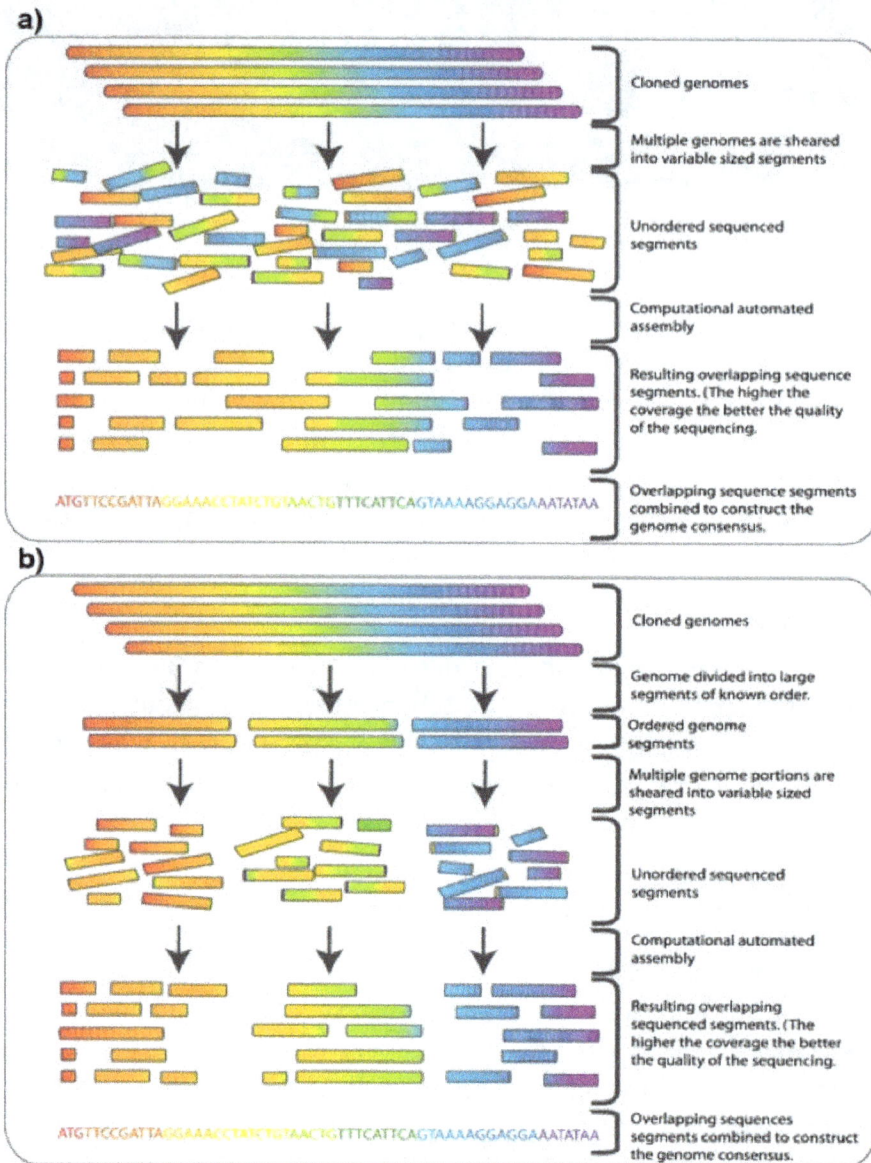

Shotgun Sequencing. By Commins, J., Toft, C., Fares, M. A. [CC BY-SA 2.5], via Wikimedia Commons

THIRD- AND FOURTH-GENERATION DNA SEQUENCING

Third-generation DNA sequencing involves the sequencing of single molecules of DNA. Fourth-generation DNA sequencing, also known as "post light sequencing," utilizes methods other than optical detection for sequencing.

BIOINFORMATICS

After sequencing, it is time to make sense of the information. The field of bioinformatics combines many fields together (i.e. biology, computer science, statistics) to use the power of computers to analyze information contained in the genomic sequence. Locating specific genes within a genome is referred to as genome annotation.

OPEN READING FRAMES (ORFS)

An open reading frame or ORF denotes a possible protein-coding gene. For double-stranded DNA, there are six reading frames to be analyzed, since the DNA is read in sets of three bases at a time and there are two strands of DNA. An ORF typically has at least 100 codons before a stop codon, with 3' terminator sequences. A functional ORF is one that is actually used by the organism to encode a protein. Computers are used to search the DNA sequence looking for ORFs, with those presumed to encode protein further analyzed by a bioinformaticist.

It is often helpful for the sequence to be compared against a database of sequences coding for known proteins. GenBank is a database of over 200 billion base pairs of sequences that scientists can access, to try and find matches to the sequence of interest. The database search tool BLAST (basic local alignment search tool) has programs for comparing both nucleotide sequences and amino acid sequences, providing a ranking of results in order of decreasing similarity.

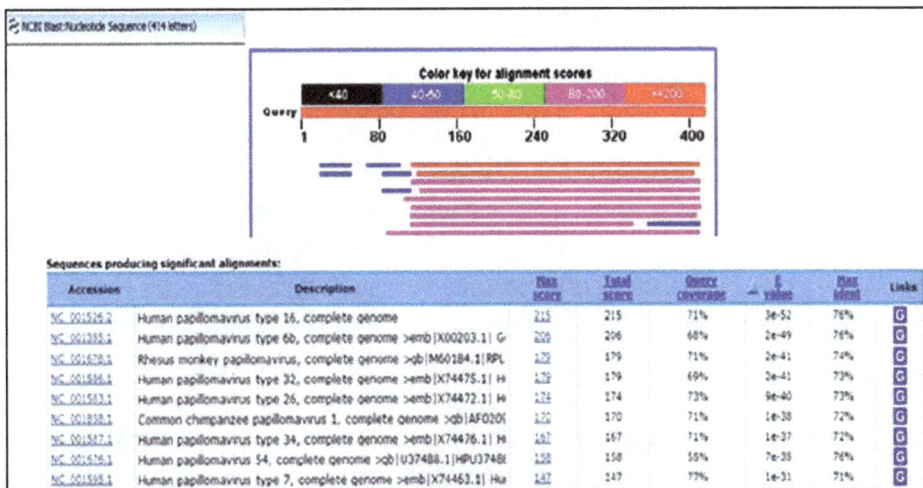

BLAST Results.

COMPARATIVE GENOMICS

Once the sequences of organisms have been obtained, meaningful information can be gathered using comparative genomics. For this genomes are assessed for information regarding size, organization, and gene content.

Comparison of the genome of microbial strains has given scientists a better picture regarding the genes that organisms pick up. A group of multiple strains share a core genome, genes coding for essential cellular functions that they all have in common. The pan genome represents all the genes found in all the members of species, so provides a good idea of the diversity of a group. Most of these "extra" genes are probably picked up by horizontal gene transfer.

Comparative genomics also shows that many genes are derived as a result of gene duplication. Genes within a single organism that likely came about because of gene duplication are referred to as paralogs. In many cases one of the genes might be altered to take on a new function. It is also possible for gene duplication to be found in different organisms, as a result of acquiring the original gene from a common ancestor. These genes are called orthologs.

FUNCTIONAL GENOMICS

The sequence of a genome and the location of genes provide part of the picture, but in order to fully understand an organism we need an idea of what the cell is doing with its genes. In other words, what happens when the genes are expressed? This is where functional genomics comes in – placing the genomic information in context.

The first step in gene expression is transcription or the manufacture of RNA. Transcriptome refers to the entire complement of RNA that a cell can make from its genome, while proteome refers to all the proteins encoded by an organisms' genome, in the final step of gene expression.

MICROARRAYS

Microarrays or gene chips are solid supports upon which multiple spots of DNA are placed, in a grid-like fashion. Each spot of DNA represents a single gene or ORF. Known fragments of nucleic acid are labeled and used as probes, with a signal produced if binding occurs. Microarrays can be used to determine what genes might be turned on or off under particular conditions, such as comparing the growth of a bacterial pathogen inside the host versus outside of the host.

PROTEOMICS

The study of the proteins of an organism (or the proteome) is referred to as proteomics. Much of the interest focuses on functional proteomics, which examines the functions of the cellular proteins and the ways in which they interact with one another.

One common technique used in the study of proteins is two-dimensional gel electrophoresis, which first separates proteins based on their isoelectric points. This is accomplished by using a pH gradient, which separates the proteins based on their amino acid content. The separated proteins are then run through a polyacrylamide gel, providing the second dimension as proteins are separated by size.

Structural proteomics focuses on the three-dimensional structure of proteins, which is often determined by protein modeling, using computer algorithms to predict the most likely folding of the protein based on amino acid information and known protein patterns.

METABOLOMICS

Metabolomics strives to identify the complete set of metabolic intermediates produced by an organism. This can be extremely complicated, since many metabolites are used by cells in multiple pathways.

METAGENOMICS

Metagenomics or environmental genomics refers to the extraction of pooled DNA directly from a specific environment, without the initial isolation and identification of organisms within that environment. Since many microbial species are difficult to culture in the laboratory, studying the metagenome of an environment allows scientists to consider all organisms that might be present. Taxa can even be identified in the absence of organism isolation using nucleic acid sequences alone, where the taxon is known as phylotype.

KEY WORDS

Genomics, sequencing, shotgun sequencing, genomic library, second generation DNA sequencing, massively parallel methods, third- and fourth-generation DNA sequencing, bioinformatics, genome annotation, open reading frame/ORF, functional ORF, GenBank, BLAST/basic local alignment search tool, comparative genomics, core genome, pan genome, paralog, ortholog, functional genomics, transcriptome, proteome, microarray/gene chips, probe, proteomics, functional proteomics, two-dimensional gel electrophoresis, structural proteomics, metabolomics, metagenomics/environmental genomics, metagenome, phylotype.

20 | The Symbioses of Microbes

INTRODUCTION

Symbiosis, strictly defined, refers to an intimate relationship between two organisms. Although many people use the term to describe a relationship beneficial to both participants, the term itself is not that specific. The relationship could be good, bad, or neutral for either partner. A mutualistic relationship is one in which both partners benefit, while a commensalistic relationship benefits one partner but not the other. In a pathogenic relationship, one partner benefits at the expense of the other. This chapter looks at a few examples of symbiosis, where microbes are one of the partners.

THE HUMAN MICROBIOME

The human microbiome describes the genes associated with all the microbes that live in and on a human. All 14^{10} of them! The microbes are mostly bacteria but can include archaea, fungi, and eukartyotic microbes The locations include skin, upper respiratory tract, stomach, intestines, and urogenital tracts. Colonization occurs soon after birth, as infants acquire microbes from people, surfaces and objects that they come in contact with.

GUT MICROBES AND HUMAN METABOLISM

Most of the microbes associated with the human body are found in the gut, particularly about 1-4 hours after eating a meal when the microbial population dramatically increases. The gut microbiota is extremely diverse and it has been estimated that from 500-1000 species of bacteria live in the human gastrointestinal tract (typically described as from 5-8 pounds of bacteria!).

The gut microbes are essential for host digestion and nutrition, aiding in digestion by breaking down carbohydrates that humans could not break down on their own, by liberating short chain fatty acids from indigestible dietary fibers. In addition, they produce vitamins such as biotin and vitamin K.

GUT MICROBES AND OBESITY

There has been increased interest in the microbial gut population, due to the possibility that it might play a role in obesity. Although currently hypothetical, research has shown that obese mice have a microbial gut community that differs from the microbes found in the gut of non-obese mice, with more Firmicutes bacteria and methanogenic Archaea. It has been suggested that these microbes are more efficient at absorbing nutrients.

HUMAN MICROBIOME AND DISEASE

It has been shown that microbiota changes are associated with diseased states or dysbiosis. Preliminary research has shown that the microbiota might be associated rheumatoid arthritis, colorectal cancer, diabetes, in addition to obesity.

RESEARCH

The Human Microbiome Project (HMP) was an international research program based in the U.S. that was focused on the functions of gut microbiota. Some 200 researchers used advanced DNA-sequencing techniques to determine what microbes are present and in what populations.

Many current research projects are focused on determining the role of the human microbiome in both health and the diseased state. There is no doubt that our knowledge will continue to grow as we find out more about the vast populations of microbes that live in and on us.

BIOFILMS

Biofilms are a complex aggregation of cells that are encased within an excellular matrix and attached to a surface. Bioforms can form on just about any surface and are common in nature and industry, being found on the surfaces of rocks, caves, pipes, boat hulls, cooking vessels, and medical implants, just to name a few. They have also been around a long time, since the fossil record shows evidence for biofilms going back 3.4 billion years!

The microbial community of a biofilm can be composed of one or two species but more commonly contains many different species of bacteria, each influencing the others gene expression and growth.

BIOFILM DEVELOPMENT

The basic steps for biofilm formation can be broken down into four steps:

- ◉ Cell disposition and attachment – in order for biofilm development to occur, free-floating or planktonic cells must collide with a suitable surface. Typically the surface has been preconditioned with the deposits of environmental proteins and other molecules.

- ◉ Colonization – cell-to-cell signaling occurs, leading to the expression of biofilm specific genes. These genes are associated with the communal production of extracellular polymeric substances. DNA released by some cells can be taken up by others, stimulating the expression of new genes.

- ◉ Maturation – the EPS matrix fully encases all the cells, as the biofilm continues to thicken and grow, forming a complex, dynamic community. Water channels form throughout the structure.

- ◉ Detachment and sloughing – individual cells or pieces of the biofilm are released to the environment, as a form of active dispersal. This release can be trigger by environmental factors, such as the concentration of nutrients or oxygen.

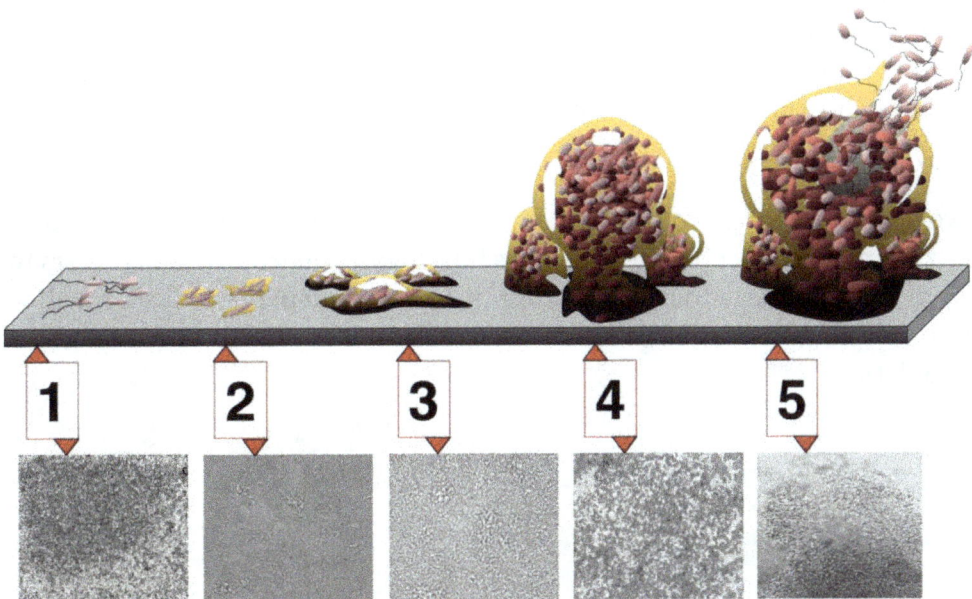

Biofilm Development. Each stage of development in the diagram is paired with a photomicrograph of a developing Pseudomonas aeruginosa biofilm. All photomicrographs are shown to same scale. By D. Davis [CC BY 2.5],

CELLULAR ADVANTAGES OF BIOFILMS

Why do bioforms develop? There are certain advantages that cells enjoy while in a biofilm, over their planktonic growth. Perhaps most importantly, biofilms offer cells increased protection from harmful conditions or substances, such as UV light, physical agitation, antimicrobial agents, and phagocytosis. It has been shown that bacteria within a biofilm are up to a thousand times more resistant to antibiotics than free-floating cells!

A biofilm also allows a cell population to "put down roots," so to speak, so that they can stay in close proximity to a nutrient-rich area. For example, a biofilm that develops on a conduit pipe at a dairy plant will have continual access to fresh food, which is much better than being swept away with the final product.

Lastly, biofilms allow for cells to grow in microbial populations, where they can easily benefit from cell-to-cell communication and genetic exchange.

BIOFILM IMPACTS

Biofilms have huge impacts throughout many different types of industry. Medical implants ranging from catheters to artificial joints are particularly susceptible to biofilm formation, leading to huge problems for the medical industry. Biofilms are responsible for many chronic infections, due to their increased resistance to antimicrobial compounds and antibiotics. A type of biofilm that affects almost everyone is the formation of dental plaque, which can lead to cavity formation.

Outside of medicine, biofilms affect just about any industry relying on pipes to convey water, food, oil, or other liquids, where their resistance makes its particularly difficult to completely eliminate the biofilm.

QUORUM SENSING

The word quorum refers to having a minimum number of members needed for an organization to conduct business, such as hold a vote. Quorum sensing refers to the ability of some bacteria to communicate in a density-dependent fashion, allowing them to delay the activation of specific genes until it is the most advantageous for the population.

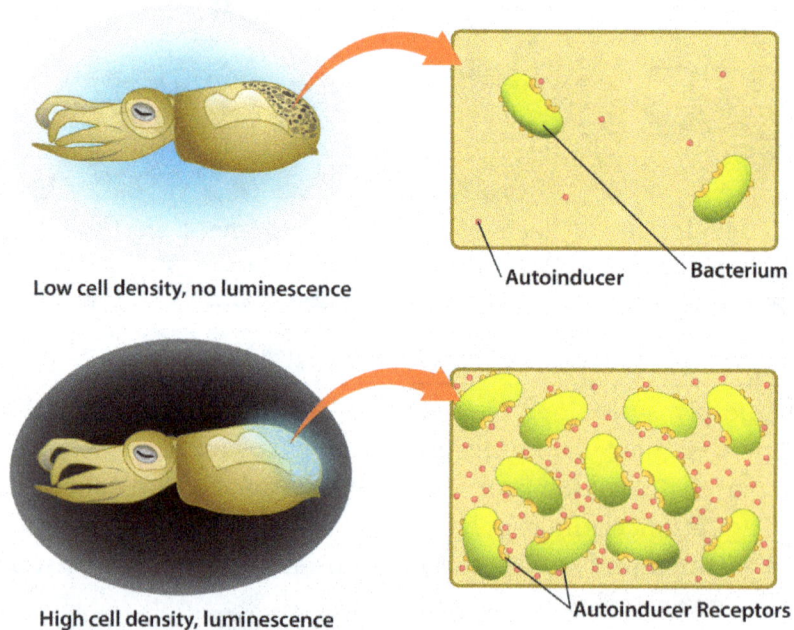

Low cell density, no luminescence

Autoinducer　　Bacterium

High cell density, luminescence

Autoinducer Receptors

Quorum sensing in Bioluminescent Bacteria.

Quorum sensing involves cell-to-cell communication, using small diffusible substances known as autoinducers. An autoinducer is produced by a cell, diffusing across the plasma membrane to be released into the environment. As the cell population increases in the environment the concentration of autoinducer increases as well, causing the molecule to bind to specific cellular receptors once a threshold concentration has been reached. The autoinducer then diffuses into the cell, often binding to a specific transcription factor. This produces a conformational change that allows the transcription factor to bind to the cell's DNA, triggering expression of specific genes.

QUORUM SENSING EXAMPLE

One of the best studied examples of quorum sensing is the mutualistic relationship between the bioluminescent bacterium Aliivibrio fischeri and the bobtail squid. The bobtail squid actually has a light organ that evolved to house the bacterium, relying on its luminescence to provide a camouflage effect against predators. At low cell density the luminescence would not provide the desired effect, representing a waste in energy by the bacterial population. Therefore, quorum sensing is used so that the lux gene that codes for the luciferase enzyme necessary for luminescence is only activated when the bacterial population is at sufficient density.

Key Words
Symbiosis, mutualistic, commensalistic, pathogenic relationship, human microbiome, dysbiosis, Human Microbiome Project (HMP), biofilm, planktonic, extracellular polymeric substances/EPS, quorum sensing, autoinducer.

21 | Introduction to Bacterial Pathogenicity

A microbe that is capable of causing disease is referred to as a pathogen, while the organism being infected is called a host. The ability to cause disease is referred to as pathogenicity, with pathogens varying in their ability. An opportunistic pathogen is a microbe that typically infects a host that is compromised in some way, either by a weakened immune system or breach to the body's natural defenses, such as a wound. The measurement of pathogenicity is called virulence, with highly virulent pathogens being more likely to cause disease in a host.

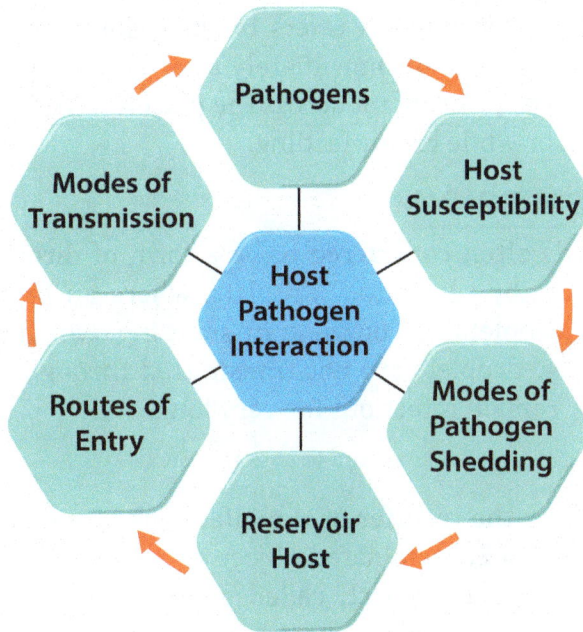

Host-Pathogen Interactions.

It is important to remember that there are many variables to take into account in a host-pathogen interaction, which is a dynamic relationship that is constantly changing. The virulence of the pathogen is important, but so is the number of microbes that gained entry to the host, the location of entry, the overall health of the host, and the state of the host's defenses. Exposure to a pathogen does not ensure that disease will occur, since a host might be able to fight off the infection before disease signs/symptoms develop.

PATHOGEN TRANSMISSION

An infection starts with exposure to a pathogen. The natural site or home for a pathogen is known as a reservoir and can either be animate (human or animal) or inanimate (water, soil, food). A pathogen can be picked up from its reservoir and then spread from one infected host to another. Carriers play an important role in the spread of disease, since they carry the pathogen but show no obvious symptoms of disease. A disease that primarily occurs within animal populations but can be spread to humans is called a zoonosis, while a hospital-acquired infection is known as a nosocomial infection.

The mechanism by which a pathogen is picked up by a host is referred to as mode of transmission, with the main mechanisms listed below:

Direct Contact

Direct contact includes host-to-host contact, such as through kissing or sexual intercourse, where one person might come in contact with another person's skin or body fluids. Vertical transmission refers to the transfer of a pathogen from mother to infant, either before or immediately after birth. An expectant mother may transmit a pathogen to her infant across the placenta while pregnant, during the act of giving birth, or while breast feeding.

DROPLET TRANSMISSION

Droplet transmission is often considered to be a form of direct contact as well. It involves transmission by respiratory droplets, where an infected host expels the pathogen in tiny droplets by coughing or sneezing, which are then inhaled by a host nearby. These droplets are not transmitted through the air over long distances, nor do they remain infectious for very long.

Indirect Contact

Indirect contact involves the transfer of the infectious agent through some type of intermediary, such as a contaminated object or person. The pathogen might be deposited on an inanimate object, called a fomite, which is then used by another person. This could include a shared toy or commonly-touched surface, like a doorknob or computer keyboard. Alternatively, a healthcare worked might transmit a pathogen from one patient to another, if they did not change their gloves between patients.

Airborne Transmission

Airborne transmission occurs due to pathogens that are in small particles or droplets in the environment, which can remain infectious over time and distance. An example might be fungal spores that are inhaled during a dust storm.

FECAL-ORAL TRANSMISSION

Fecal-oral transmission occurs when an infected host is shedding the pathogen in their feces which contaminate food or water that is consumed by the next host.

VECTORBORNE TRANSMISSION

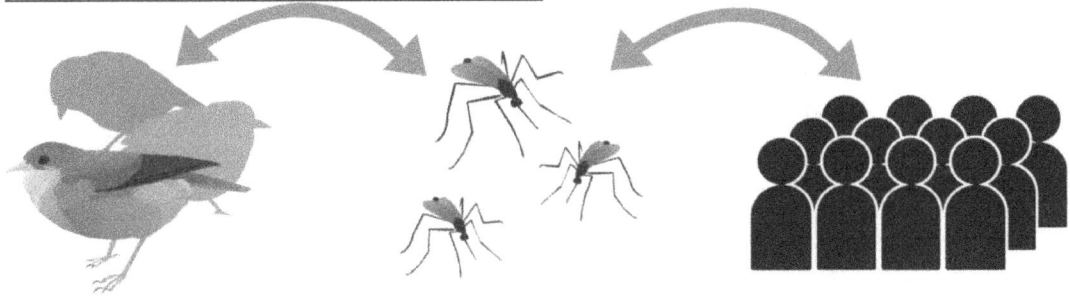

Pathogen Transmission.

Vectorborne transmission occurs when an arthropod vector, such as mosquitoes, flies, ticks, are involves in the transmission. Sometimes the vector just picks up the infectious agents on their external body parts and carries it to another host, but typically the vector picks up the infectious agent when biting an infected host. The agent is picked up in the blood, and then spread to the next host when the vector moves on to bite someone else.

VIRULENCE FACTORS

In order for a bacterium to be virulent, it must have capabilities that allow it to infect a host. These capabilities arise from physical structures that the bacterium has or chemical substances that the bacterium can produce. Collectively the characteristics that contribute to virulence are called virulence factors.

The genes that code for virulence factors are commonly found clustered on the pathogen's chromosome or plasmid DNA, called pathogenicity islands. These pathogenicity islands can be distinguished by a G+C content that differs from the rest of the genome and the presence of insertion-like sequences flanking the gene cluster. Pathogenicity islands facilitate the sharing of virulence factors between bacteria due to horizontal gene transfer, leading to the development of new pathogens over time.

Often the genes for virulence factors are controlled by quorum sensing, to ensure gene activation when the pathogen population is at an optimal density. Triggering the genes too soon could alert the host's immune system to the invader, cutting short the bacterial infection.

ADHERENCE AND COLONIZATION

Bacterial pathogens must be able to grab onto host cells or tissue, and resist removal by physical means (such as sneezing) or mechanical means (such as movement of the cilated cells that line our airway). Adherence can involve polysaccharide layers made by the bacteria, such as a capsule or slime layer, which provide adhesion to host cells as well as resistance from phagocytosis. Adherence can also be accomplished by physical structures such as a pilus or flagellum.

Once cells are successfully adhering to a surface, they increase in number, utilizing resources available at the site. This colonization is important for pathogen survival and invasion to other sites, which will yield increased nutrients and space for the growing population.

INVASION

Invasion refers to the ability of the pathogen to spread to other locations in the host, by invading host cells or tissue. It is typically at this point when disease or obvious signs/symptoms of illness will occur. While physical structures can still play a role in invasion, most bacterial pathogens produce a wide array of chemicals, specifically enzymes that effect the host's cells and tissue. Enzymes such as collagenase, which allows the pathogen to spread by breaking down the collagen found in connective tissue. Or leukocidins, which destroy the host's white blood cells, decreasing resistance. Hemolysins lyse the host's red blood cells, releasing iron, a growth-limiting factor for bacteria.

Bacteria in the bloodstream, a condition known as bacteremia, can quickly spread to locations throughout the host. This can result in a massive, systemic infection known as septicemia, which can result in septic shock and death, as the host becomes overwhelmed by the bacterial pathogen and its products.

TOXINS

Toxins are a very specific virulence factor produced by some bacterial pathogens, in the form of substances that are poisonous to the host. Toxigenicity refers to an organism's ability to make toxins. For bacteria, there are two categories of toxins, the exotoxins and the endotoxins.

EXOTOXINS

Exotoxins are heat-sensitive soluble proteins that are released into the surrounding environment by a living organism. These incredibly potent substances can spread throughout the host's body, causing damage distant from the original site of infection. Exotoxins are associated with specific diseases, with the toxin genes often carried on plasmids or by prophages. There are many different bacteria that produce exotoxins, causing diseases such as botulism, tetanus, and diphtheria.

There are three categories of exotoxins:

- ◉ Type I: cell surface-active – these toxins bind to cell receptors and stimulate cell responses. One example is superantigen, that stimulates the host's T cells, an important component of the immune system. The stimulated T cells produce an excessive amount of the signaling molecule cytokine, causing massive inflammation and tissue damage.

- ◉ Type II: membrane-damaging – these toxins exert their effect on the host cell membrane, often by forming pores in the membrane of the target cell. This can lead to cell lysis as cytoplasmic contents rush out and water rushes in, disrupting the osmotic balance of the cell.

- ◉ Type III: intracellular – these toxins gain access to a particular host cell and stimulate a reaction within the target cell. One example is the AB-toxin – these toxins are composed of two subunits, an A portion and a B portion. The B subunit is the binding portion of the toxin, responsible for recognizing and binding to the correct cell type. The A subunit is the portion with enzymatic activity. Once delivered into the correct cell by the B subunit, the A subunit enacts some mechanism on the cell, leading to decreased cell function and/or cell death. An example is the tetanus toxin produced by the bacterium Clostridrium tetani. Once delivered to a neuron, the A subunit will cleave the cellular synaptobreven, resulting in a decrease in neurotransmitter release. This results in spastic paralysis of the host. Each AB-toxin is associated with a different disease.

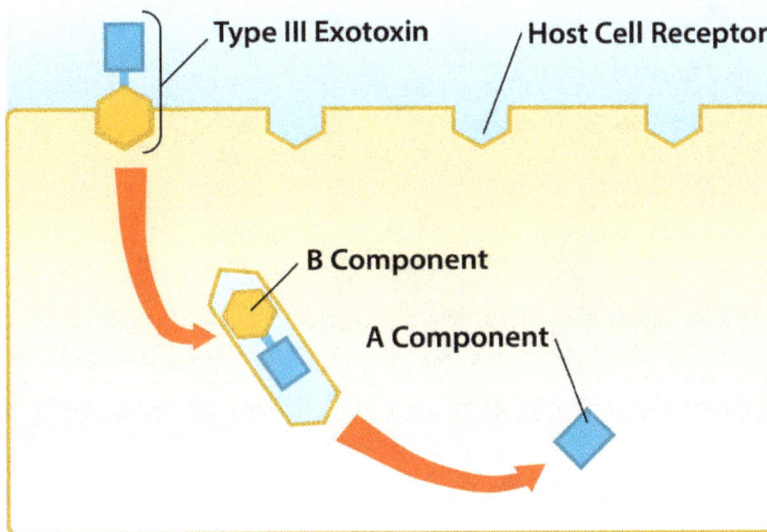

AB-toxin: Host Cell Binding.

ENDOTOXINS

Endotoxins are made by gram negative bacteria, as a component of the outer membrane of their cell wall. The outer membrane contains lipopolysaccharide or LPS, with the toxic component being the lipid part known as lipid A. Lipid A is heat-stable and is only released when the bacterial cell is lysed. The effect on the host is the same, regardless of what bacterium made the lipid A – fever, diarrhea, weakness, and blood coagulation. A massive release of endotoxin in a host can cause endotoxin shock, which can be deadly.

Key Words
Pathogen, host, pathogenicity, opportunistic pathogen, virulence, reservoir, carrier, nosocomial infection, mode of transmission, direct contact, vertical contact, droplet transmission, indirect contact, fomite, airborne transmission, fecal-oral transmission, vectorborne transmission, virulence factor, pathogenicity island, adherence, colonization, invasion, bacteremia, septicemia, toxin, toxigenicity, exotoxin, Type I/cell surface-active toxin, superantigen, T cell, cytokine, Type II/membrane-damaging toxin, Type III/intracellular toxin, AB toxin, endotoxin, lipid A, endotoxin shock.

VIRAL CLASSIFICATION

Since viruses lack ribosomes (and thus rRNA), they cannot be classified within the Three Domain Classification scheme with cellular organisms. Alternatively, Dr. David Baltimore derived a viral classification scheme, one that focuses on the relationship between a viral genome to how it produces its mRNA. The Baltimore Scheme recognizes seven classes of viruses.

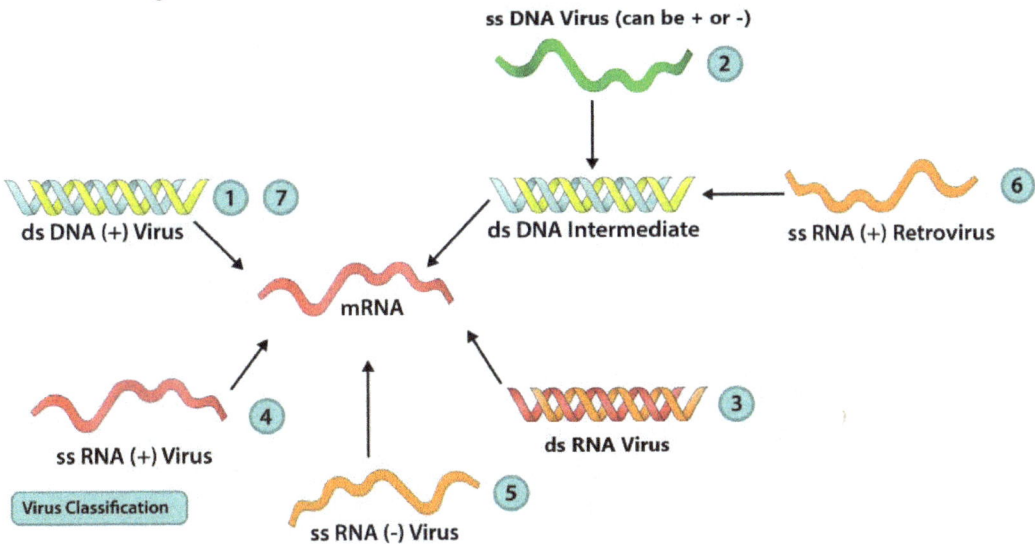

DNA Viruses

Class I: dsDNA

DNA viruses with a dsDNA genome, like bacteriophages T4 and lambda, have a genome exactly the same as the host cell that they are infecting. For this reason, many host enzymes can be utilized for replication and/or protein production. The flow of information follows a conventional pathway: dsDNA → mRNA → protein, with a DNA-dependent RNA-polymerase producing the mRNA and the host ribosome producing the protein. The genome replication, dsDNA → dsDNA, requires a DNA-dependent DNA-polymerase from either the virus or the host cell.

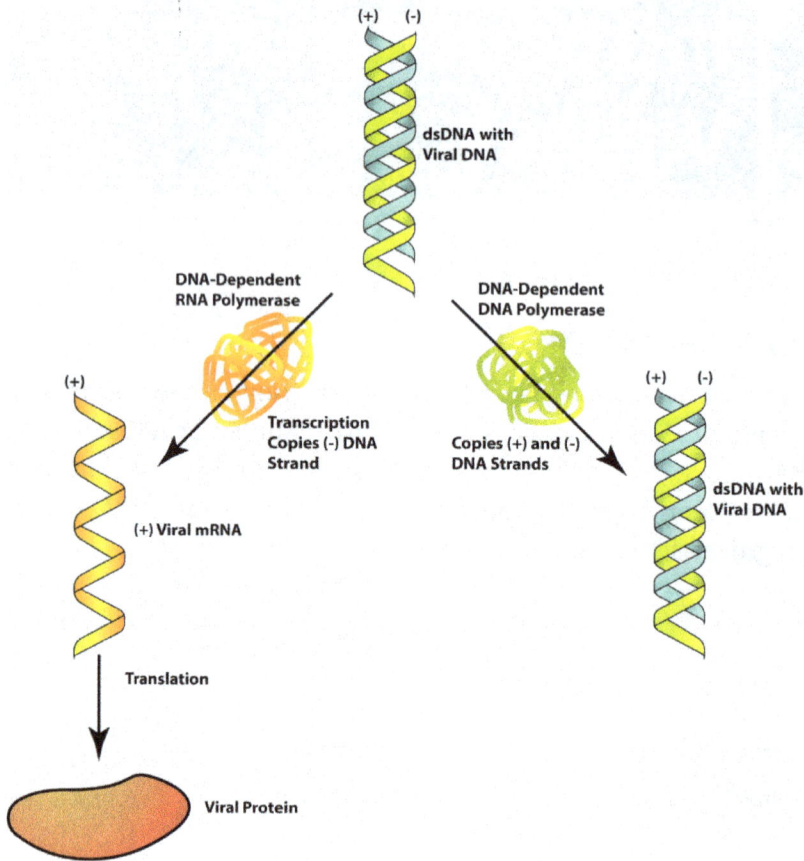

(+) **(-)**

**dsDNA with
Viral DNA**

**DNA-Dependent
RNA Polymerase**

**DNA-Dependent
DNA Polymerase**

(+)

**Transcription
Copies (-) DNA
Strand**

**Copies (+) and (-)
DNA Strands**

(+) **(-)**

**dsDNA with
Viral DNA**

(+) Viral mRNA

Translation

Viral Protein

dsDNA.

The virus often employs strategies for control of gene expression, to insure that particular viral products are made at specific times in the virus replication. In the case of T4, the host RNA polymerase binds to the viral DNA and begins transcribing early genes immediately after the DNA is injected into the cell. One of the early viral proteins modifies the host RNA polymerase so that it will no longer recognize host promoters at all, in addition to moving on to transcribe genes for middle-stage viral proteins. A further modification (catalyzed by middle-stage viral proteins) further modified the RNA polymerase so it will recognize viral genes coding for late-stage proteins. This insures an orderly production of viral proteins.

The replication of several dsDNA viruses results in the production of concatemers, where several viral genomes are linked together due to short single-stranded regions with terminal repeats. As the genome is packaged into the capsid a viral endonuclease cuts the concatemer to an appropriate length.

There are several animal viruses with dsDNA genomes, such as the pox viruses and the adenoviruses. The herpesviruses have several notable features, such as the link of several members with cancer and the ability of the viruses to remain in a latent form within their host. A productive infection results in an explosive viral population, cell death, and development of disease signs, during which neurons are infected. A latent infection develops in the neurons, allowing the virus to remain undetected in the host. If the viral genome is reactivated, a productive infection results, leading to viral replication and disease signs again.

CLASS II: ssDNA

The flow of information for ssDNA viruses, such as the parvoviruses, will still follow the conventional pathway, to a certain extent: DNA → mRNA → protein. But the viral genome can either have the same base sequence as the mRNA (plus-strand DNA) or be complementary to the mRNA (minus-strand DNA). In the former case, a DNA strand that is complementary to the viral genome must be manufactured first, forming a double-stranded replicative form (RF). This can be used to both manufacture viral proteins and as a template for viral genome copies. For the minus-strand DNA viruses, the genome can be used directly to produce mRNA but a complementary copy will still need to be made, to serve as a template for viral genome copies.

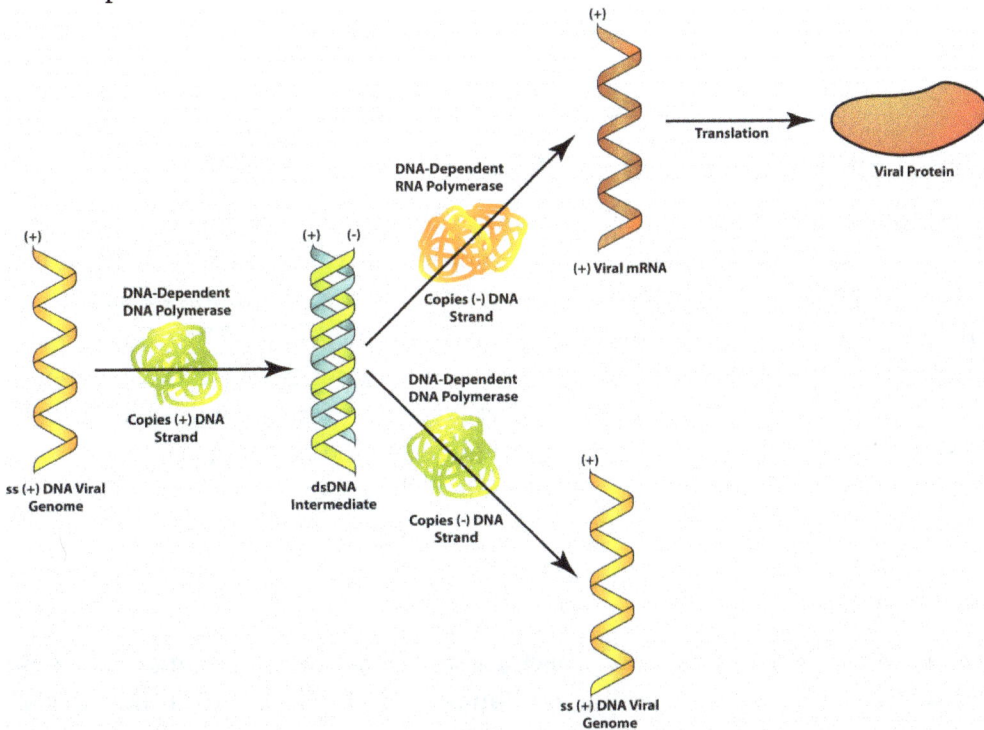

ssDNA.

The replicative form can be used for rolling-circle replication, where one strand is nicked and replication enzymes are used to extend the free 3' end. As a complementary strand is synthesized around the circular DNA, the 5' end is peeled off, leading to a displaced strand that continues to grow in length.

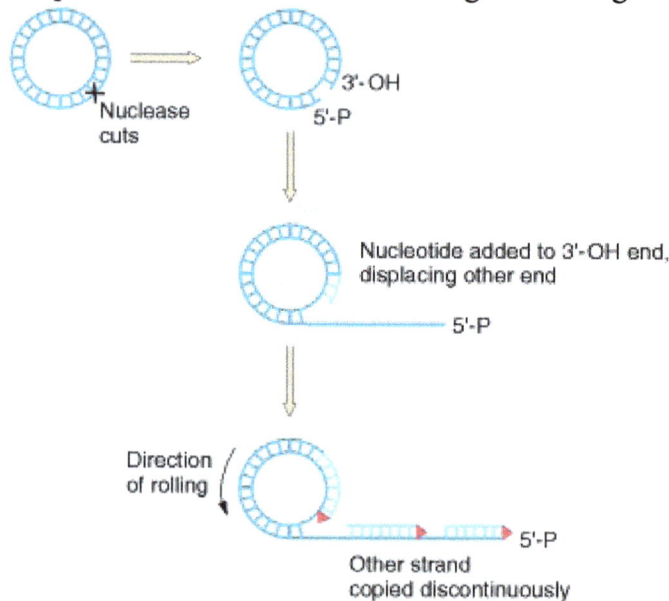

Rolling-Circle Replication.

Class VII: DNA Viruses that Use Reverse Transcriptase

The hepadnaviruses contain a DNA genome that is partially double-stranded, but contains a single-stranded region. After gaining entrance into the cell's nucleus, host cell enzymes are used to fill in the gap with complementary bases to form a dsDNA closed loop. Gene transcription yields a plus-strand RNA known as the pregenome, as well as the viral enzyme reverse transcriptase, an RNA-dependent DNA-polymerase. The pregenome is used as a template for the reverse transcriptase to produced minus-strand DNA genomes, with a small piece of pregenome used as a primer to produce the double-stranded region of the genomes.

RNA VIRUSES

Class III: dsRNA

Double-stranded RNA viruses infect bacteria, fungi, plants, and animals, such as the rotavirus that causes diarrheal illness in humans. But cells do not utilize dsRNA in any of their processes and have systems in place to destroy any dsRNA found in the cell. Thus the viral genome, in its dsRNA form, must be hidden or protected from the cell enzymes. With very few exceptions, cells also lack RNA-dependent RNA-polymerases, necessary for replication of the viral genome so the virus must provide this enzyme itself. The viral RNA-dependent RNA polymerase acts as both a transcriptase to transcribe mRNA, as well as a replicase to replicate the RNA genome.

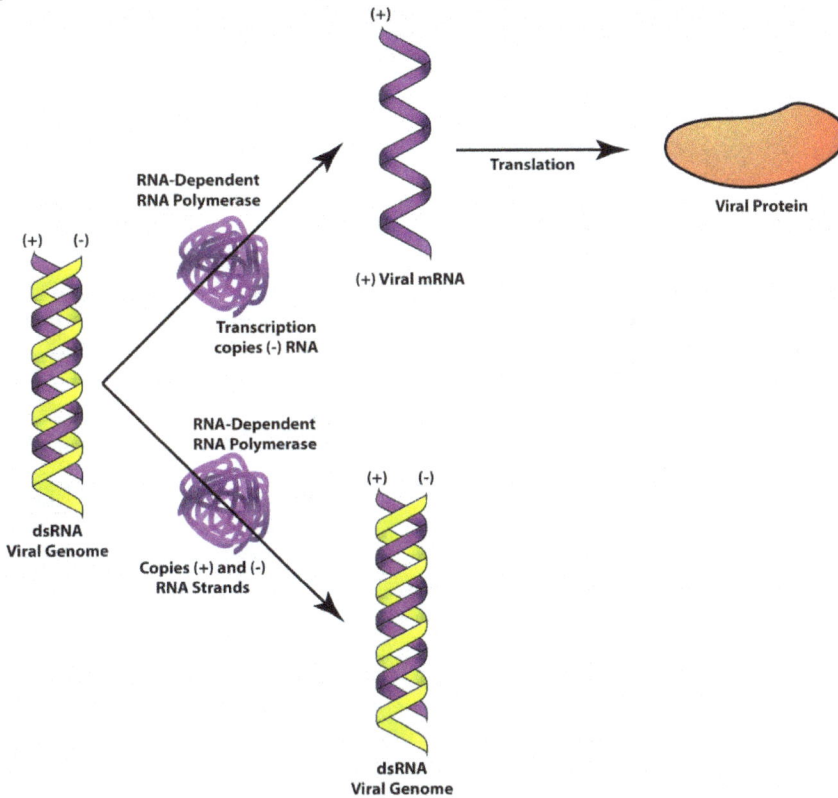

dsRNA.

For the rotavirus, the viral nucleocapsid remains intact in the cytoplasm with replication events occurring inside, allowing the dsRNA to remain protected. Messenger RNA is transcribed from the minus-strand of the RNA genome and then translated by the host ribosome in the cytoplasm. Viral proteins aggregate to form new nucleocapsids around RNA replicase and plus-strand RNA. The minus-strand RNA is then synthesized by the RNA replicase within the nucleocapsid, once again insuring protection of the dsRNA genome.

Class IV: +ssRNA

Viruses with plus-strand RNA, such as poliovirus, can use their genome directly as mRNA with translation by the host ribosome occurring as soon as the unsegmented viral genome gains entry into the cell. One of the viral genes expressed yields an RNA-dependent RNA-polymerase (or RNA replicase), which creates minus-strand RNA from the plus-strand genome. The minus-strand RNA can be used as a template for more plus-strand RNA, which can be used as mRNA or as genomes for the newly forming viruses.

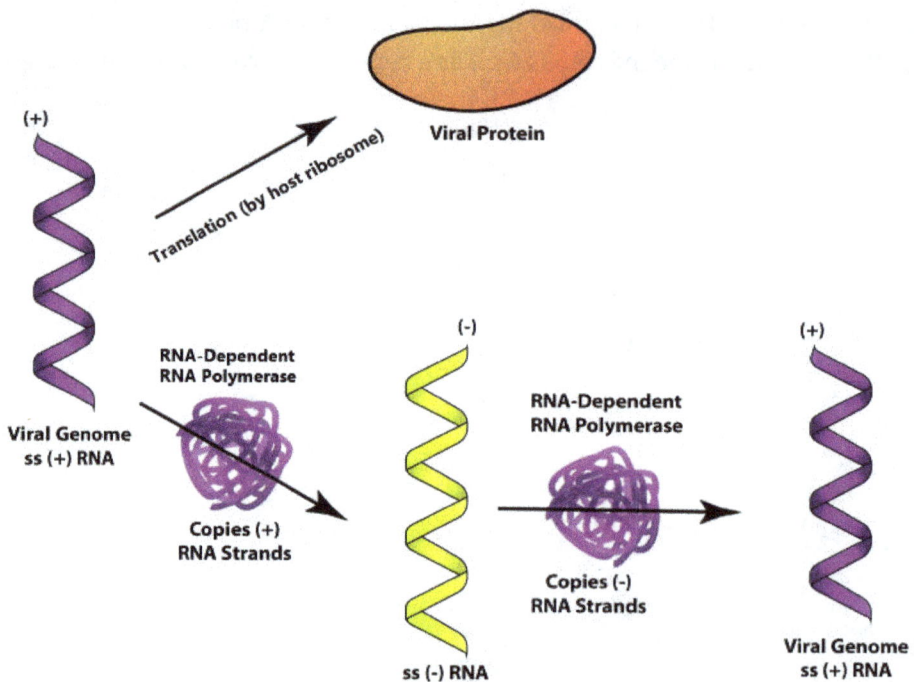

Translation of the poliovirus genome yields a polyprotein, a large protein with protease activity that cleaves itself into three smaller proteins. Additional cleavage activity eventually yields all the proteins needed for capsid formation, as well as an RNA-dependent RNA-polymerase.

The formation of a polyprotein that is cut into several smaller proteins illustrates one possible strategy to an issue faced by many +ssRNA viruses – how to generate multiple proteins from an unsegmented +ssRNA genome? Other possibilities include:

- subgenomic mRNA – during translation, portions of the viral RNA may be skipped, resulting in different proteins than what is made from the viral RNA in its entirety.

- ribosomal frame-shifting – the ribosome "reads" the mRNA in groups of three nucleotides or codon, which translate to one amino acid. If the ribosome starts with nucleotide #1, that is one open reading frame (ORF), resulting in one set of amino acids. If the ribosome were to move forward where nucleotide 2 is the starting nucleotide that would be ORF #2, resulting in a completely different set of amino acids. If the ribosome were to move forward again where nucleotide 3 is the starting nucleotide that would be ORF#3, resulting in an entirely different set of amino acids. Some viruses have viral genes that deliberately overlap within different ORFs, leading to the production of different proteins from a single mRNA.

- readthrough mechanism – a viral genome can have stop codons embedded throughout the sequence. When the ribosome comes to a stop codon it can either stop, ending the amino acid sequence, or it can ignore the stop codon, continuing on to make a longer string of amino acids. For viruses with the readthrough mechanism, they acquire a variety of proteins by having stop codons that are periodically ignored. Sometimes this function is combined with the ribosomal frame-shifting to produce an even greater variety of viral proteins.

Class V: -ssRNA

Minus-strand RNA viruses include many members notable for humans, such as influenza virus, rabies virus, and Ebola virus. Since the genome of minus-strand RNA viruses cannot be used directly as mRNA, the virus must carry an RNA-dependent RNA-polymerase within its capsid. Upon entrance into the host cell, the plus-strand RNAs generated by the polymerase are used as mRNA for protein production. When viral genomes are needed the plus-strand RNAs are used as templates to make minus-strand RNA.

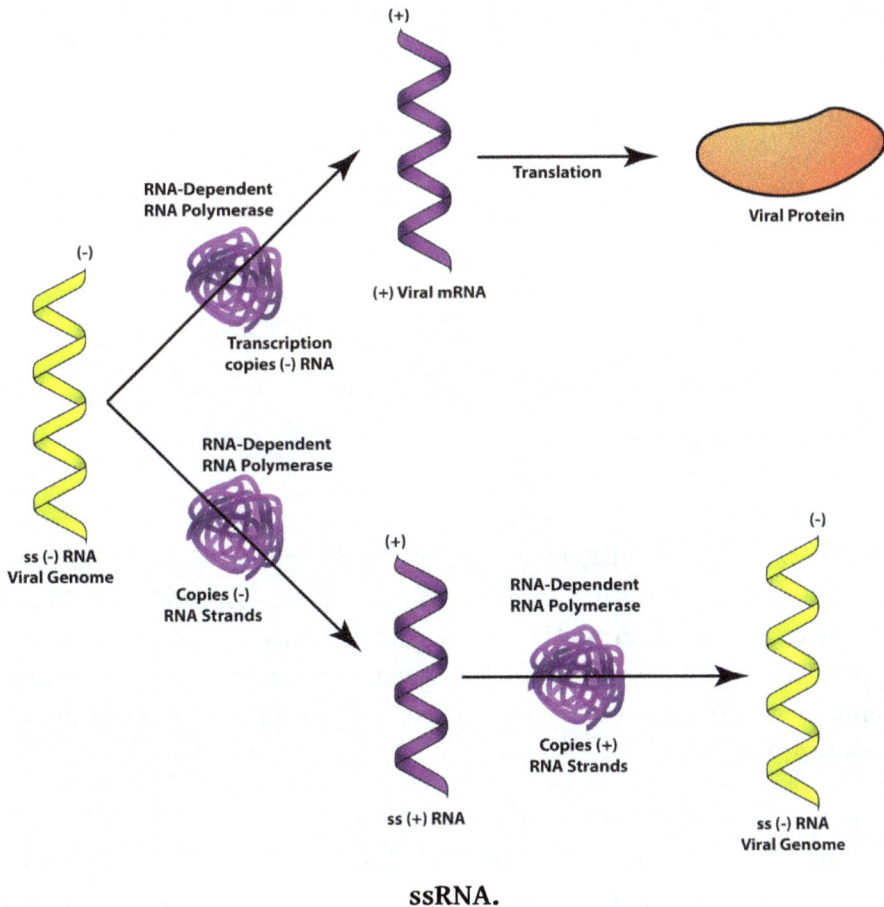

ssRNA.

Class VI: +ssRNA, retroviruses

Despite the fact that the retroviral genome is composed of +ssRNA, it is not used as mRNA. Instead, the virus uses its reverse transcriptase to synthesize a piece of ssDNA complementary to the viral genome. The reverse transcriptase also possesses ribonuclease activity, which is used to degrade the RNA strand of the RNA-DNA hybrid. Lastly, the reverse transcriptase is used as a DNA polymerase to make a complementary copy to the ssDNA, yielding a dsDNA molecule. This allows the virus to insert its genome, in a dsDNA form, into the host chromosome, forming a provirus. Unlike a prophage, a provirus can remain latent indefinitely or cause the expression of viral genes, leading to the production of new viruses. Excision of the provirus does not occur for gene expression.

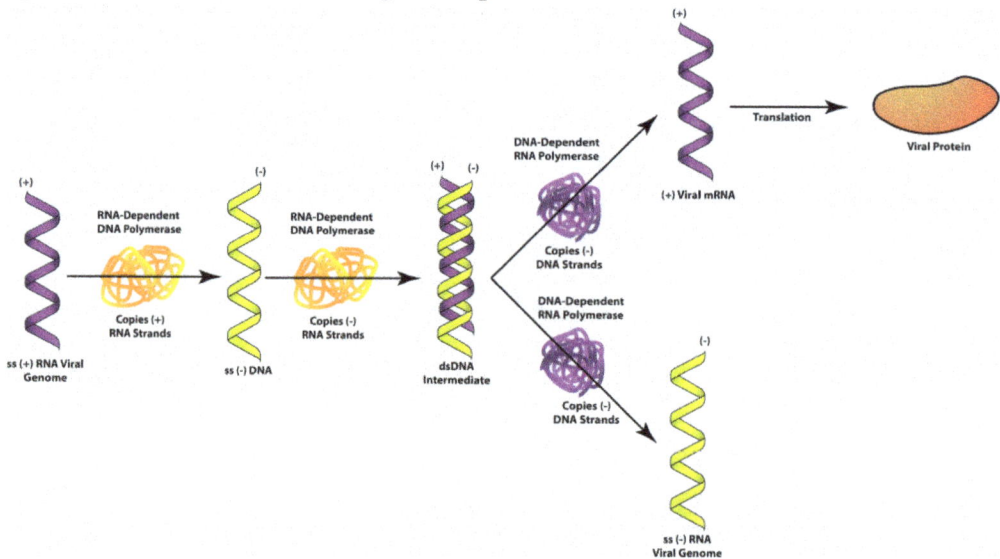

+ssRNA, retroviruses.

OTHER INFECTIOUS AGENTS

Viroids

Viroids are small, circular ssRNA molecules that lack protein. These infectious molecules are associated with a number of plant diseases. Since ssRNA is highly susceptible to enzymatic degradation, the viroid RNA has extensive complementary base pairing, causing the viroid to take on a hairpin configuration that is resistant to enzymes. For replication viroids rely on a plant RNA polymerase with RNA replicase activity.

Prions

Prions are infectious agents that completely lack nucleic acid of any kind, being made entirely of protein. They are associated with a variety of diseases, primarily in animals, although a prion has been found that infects yeast (!). Diseases include bovine spongiform encephalopathy (BSE or "mad cow disease"), Creutzfeld-Jakob disease in humans, and scrapie in sheep.

The prion protein is found in the neurons of healthy animals (PrPC or Prion Protein Cellular), with a particular secondary structure. The pathogenic form (PrPSC or Prion Protein Scrapie) has a different secondary structure and is capable of converting the PrPC into the pathogenic form. Accumulation of the pathogenic form causes destruction of brain and nervous tissue, leading to disease symptoms such as memory loss, lack of coordination, and eventually death.

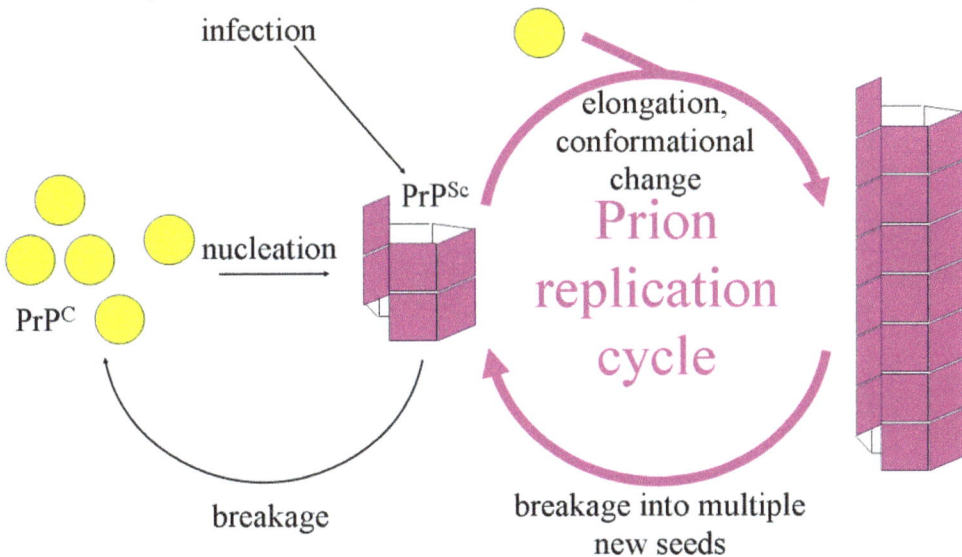

Prions. Joannamasel at English Wikipedia [CC BY-SA 3.0], via Wikimedia Commons

Key Words

Baltimore Scheme, Class I, Class II, Class III, Class IV, Class V, Class VI, Class VII, DNA-dependent RNA polymerase, DNA-dependent DNA-polymerase, concatemer, productive infection, latent infection, plus-strand DNA/+DNA, minus-strand DNA/-DNA, dsDNA, ssDNA, replicative form (RF), rolling-circle replication, pregenome, reverse transcriptase, RNA-dependent DNA-polymerase, dsRNA, RNA-dependent RNA-polymerase, transcriptase, replicase, plus-strand RNA/+ssRNA, minus-strand RNA/-ssRNA, polyprotein, subgenomic mRNA, ribosomal frame-shifting, open reading frame (ORF), readthrough mechanism, stop codon, retrovirus, ribonuclease, provirus, viroid, prion, PrPC/Prion Protein Cellular, PrPSC/Prion Protein Scrapie.

Index

www.ingramcontent.com/pod-product-compliance
Lightning Source LLC
Chambersburg PA
CBHW062008190326
41458CB00009B/3007

* 9 7 8 1 7 8 7 1 5 1 3 4 5 *